CAMBRIDGE TEXTS IN
THE PHYSIOLOGICAL SCIENCES

Editors: R. S. COMLINE, A. W. CUTHBERT
K. C. DIXON, J. HERBERT
S. D. IVERSEN, R. D. KEYNES
H. L. KORNBERG

4   Energy balance and temperature regulation

Other titles in this series

Energy balance and temperature regulation

M. W. STANIER

NEWNHAM COLLEGE CAMBRIDGE

L. E. MOUNT

CLIMATIC RESEARCH UNIT
UNIVERSITY OF EAST ANGLIA

and

J. BLIGH

INSTITUTE OF ARCTIC BIOLOGY UNIVERSITY OF ALASKA

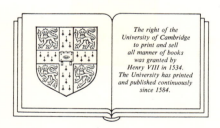

The right of the
University of Cambridge
to print and sell
all manner of books
was granted by
Henry VIII in 1534.
The University has printed
and published continuously
since 1584.

CAMBRIDGE UNIVERSITY PRESS

CAMBRIDGE

LONDON   NEW YORK   NEW ROCHELLE

MELBOURNE   SYDNEY

Published by the Press Syndicate of the University of Cambridge
The Pitt Building, Trumpington Street, Cambridge, CB2 1RP
32 East 57th Street, New York, NY 10022, USA
296 Beaconsfield Parade, Middle Park, Melbourne 3206, Australia

First published 1984

Printed in Great Britain by the University Press, Cambridge

Library of Congress catalogue card number: 83–18950

British Library cataloguing in publication data
Stanier, M.W.
Energy balance and temperature regulation.
1. Energy metabolism
I. Title II. Mount, L.E. III. Bligh, J.
59.1′3′3   QP171

ISBN 0 521 25827 8 hard covers
ISBN 0 521 27727 2 paperback

Contents

Preface

This book is intended for first- and second-year undergraduates studying animal physiology in connection with courses in Biology or Environmental Sciences, Medical or Veterinary Sciences. It deals with a particular aspect of the physiology of homeothermic vertebrates – man, other mammals, and birds: that is, the energy metabolism of such species and their exchange of energy with the environment. Since a large part of the homeotherm's metabolism is concerned with the maintenance of a stable body temperature, the first chapters show how this is done, describing the various factors which influence heat production within the body, heat loss from it, and the barriers to this heat loss. The normal body temperature and the limits of the normal range in various homeotherms are described, and the central and peripheral mechanisms controlling the body temperature are discussed. The particular problems of temperature control for the new-born and in old age are considered. The mechanisms giving the homeotherm the ability to colonize or to move between different climates are reviewed, and the assessment of the climatic conditions for comfort in humans is described. A later chapter deals with energy intake and retention, the storage of energy, the alteration of body composition and the problem of obesity in relation to metabolic rate. Two special aspects of energy metabolism – muscular exercise and fever – are discussed in the two final chapters.

The authors are grateful to Dr D. L. Ingram and Dr W. H. Close who have read and commented on the whole or parts of the book in typescript. They also thank the following individuals and publishers for permission to reproduce published or unpublished figures:
Drs G. Alexander, J. Aschoff, J. Bassey, K. Cena, R. P. Clark, W. H. Close, K. J. Collins, M. J. Dauncey, O. G. Edholm, G. E. Folk, E. N. Hey, D. L. Ingram, H. W. Newlands, J. D. Pullar, and B. P. Setchell; Academic Press, American Society of Animal Science, Annual Reviews Inc., Archives of Disease in Childhood, British Medical Bulletin, British Medical Journal, Cambridge University Press, Edward Arnold, Granada Publishing Co., Lancet, Lea & Febiger, Springer New York, Springer-Verlag.

CHAPTER ONE

Heat balance, metabolic rate and body temperature

The total sum of matter and energy in the observable universe appears to be constant, but its distribution is constantly changing. The changes which occur spontaneously are those which make the distribution of matter and energy more random, or, as a physical chemist would say, which increase entropy. However some changes do occur which make for increased order and organization of matter at the expense of increased degradation of energy; the bodies of living creatures, plant or animal, are an example of this.

Numerous chemical reactions are involved in the metabolism (growth, maintenance, and repair) of living organisms. These reactions change chemical energy from one form to another, and some energy appears as heat. The heat thus generated is very soon dissipated from the surface of the body of most species of animals, so the body temperature does not rise, and the animal is a 'temperature conformer': its body temperature is identical to that of its surroundings, whether ocean, freshwater or air. Two groups of animals that evolved relatively recently, the birds and the mammals, are able to maintain a fairly stable body temperature which is normally well above that of the surrounding water or air. These are 'temperature regulators' which produce much more metabolic heat than the 'conformers'. 'Conformers' and 'regulators' are also known respectively as poikilothermic ('varied heat') and homeothermic ('similar heat'). (The vernacular terms 'cold-blooded' and 'warm-blooded' are now obsolete in scientific literature.) Although this book is concerned only with birds and mammals (including man), it must always be remembered that these are a small minority of all animal species, and that some general principles, such as the equality of energy intake with output plus storage, apply to all species – poikilotherms and homeotherms, animals and plants.

THE Q_{10} EFFECT

There are certain advantages in maintaining a constant body temperature. The body of the homeotherm is able to maintain a fairly constant degree of activity, summer and winter, day and night. Again, many homeotherms are able to move about on the earth's surface and colonize the tropics and the polar regions. If the internal environment (the fluid surroundings) of all living cells of the animal's body is kept at a uniform temperature, the chemical reactions within the cell proceed at a fairly constant rate; a lower temperature reduces the rate, a higher temperature increases it. It has been found that if a chemical reaction proceeds at a given rate at room temperature (say 20 °C), it will proceed about twice as fast if the temperature goes up by 10 °C, and twice as fast again for another 10 °C; or, put more briefly, the reaction has a Q_{10} of 2. (This phenomenon is also called the Arrhenius effect after the scientist who first described it.) Many biochemical reactions have a Q_{10} between 2 and 3. This means that a fall in body temperature to say, 5 °C, slows down the rate of all vital reactions. At the upper end of the range, there would be a limitation by the temperature at which body proteins start to denature, about 45–50 °C. Most homeotherms maintain a fairly stable body temperature between 35 °C and 42 °C depending on species and other conditions.

METABOLIC RATE

Though the life of the homeotherm has certain advantages, it is very expensive in terms of energy requirement. It has been calculated that a 10 kg mammal generates about seven times as much energy as a 10 kg lizard, even when the lizard has been put in a warm environment so that its body temperature is 37 °C, the same as that of the mammal. The total of all the chemical reactions of an animal's body is called its metabolism, and the sum of the rates of the reactions is the metabolic rate of the body. Those reactions which form new tissue or synthesize large molecules from smaller ones are called anabolic; those which break down molecules or tissues are catabolic. Some of the body's chemical reactions are endothermic, absorbing heat, but most are exothermic, liberating heat. So, if one wants to measure the total metabolism, one can measure the rate of heat loss over a certain period of time (minutes or hours). If the weight and the body temperature do not change appreciably in this time, the rate of heat loss (in kJ h^{-1} or watts) is equal to the rate at which heat is generated in the body during that time. This kind of measurement is called *direct calorimetry*. Since it happens that most of the exothermic reactions in the body of homeotherms depend, in the long run, on oxidation, a measurement of oxygen consumption can also be used as an indication of metabolic rate. This is called *indirect calorimetry*. These two methods will be described.

MEASUREMENT OF METABOLIC RATE

Direct calorimetry

Direct calorimetry for small animals or birds for short periods of time is not difficult. The animal is placed in a small metal container, immersed in a water-bath, and the heat from the animal's body, conducted through the metal, warms the water by an amount which can be measured precisely. Direct calorimetry of humans and large mammals uses air instead of water as the medium of heat-uptake. Air of a known controllable temperature is drawn through a small well-insulated room in which the subject is placed. The person's body heat warms the air, and water evaporating from the lungs and skin moistens it. The air is then drawn through a water-absorber to collect and measure the amount of water; this allows calculation of evaporative heat loss. The air then passes through a heat-exchanger where it is cooled to its original temperature before being re-circulated to the room. The work done by the heat-exchanger in cooling the air gives a measure of the person's non-evaporative heat loss. The heat-exchanger is calibrated beforehand by placing in the room a heat source of precisely known wattage. Any food eaten by the subject during the measurements must first be equilibrated at the temperature of the room; heat needed to warm this mass of food to his body temperature can be calculated. Though the method of direct human calorimetry is simple in principle, the various precautions needed for obtaining accurate and reproducible results make the apparatus quite elaborate.

Direct calorimetry of humans was performed during the early years of this century by the pioneers of the study of nutrition and metabolism: Benedict, Lusk, Dubois and others. There has recently been a revival of interest in human calorimetry, making use of the various technical advances which have occurred in the meantime. Research in this field is now concerned mainly with studies of obesity and hormonal disorders, and there are three or four human calorimeters in use in several parts of Europe. Direct calorimeters for farm animals are in use in several research centres in Europe and North America, for the purpose of studying animal production and husbandry.

Indirect calorimetry

Measurement of metabolic rate by indirect calorimetry makes use of the fact that heat is generated in the body by oxidation of a mixture of carbohydrate, fat and protein, a group of reactions which use oxygen and produce carbon dioxide and nitrogenous excretory compounds. Measurement of oxygen used by the body in a given period, and carbon dioxide and nitrogenous excreta produced, can thus give an indication of the heat generated in this time.

For oxidation of pure carbohydrate (glucose):
$$C_6H_{12}O_6 + 6O_2 = 6CO_2 + 6H_2O + 2820 \text{ kJ}$$
$$\text{Respiratory quotient (RQ)} = \frac{\text{vols. } CO_2 \text{ produced}}{\text{vols. } O_2 \text{ consumed}} = \frac{6}{6} = 1$$

For oxidation of pure fat (triolein):
$$C_{57}H_{104}O_6 + 80O_2 = 57CO_2 + 52H_2O + 32\,083 \text{ kJ}$$
$$RQ = \frac{57}{80} = 0.71.$$

No equation can be written for oxidation of protein, because of uncertainties about precise routes of breakdown, but it has been found experimentally that oxidation of 1 g of a typical protein in the body uses 966 ml oxygen and produces 774 ml carbon dioxide, and 0.16 g urinary nitrogen. One litre of oxygen used in oxidation of protein generates 18.57 kJ. The procedure for experiments lasting 24 hours or more is to measure the total oxygen intake, total carbon dioxide output, and urinary nitrogen excretion of the subject (human or other) in this time. The urinary nitrogen gives a measure of the protein metabolized, since g urinary nitrogen $\div 0.16 =$ g protein $= P$. From the total oxygen intake, $P \times 966$ is subtracted, to give the oxygen required for oxidation of carbohydrate and fat, x ml. Similarly, from the total carbon dioxide output, $P \times 774$ is subtracted, to give the output of carbon dioxide derived from oxidation of carbohydrate and fat, y ml. The *non-protein* RQ is then y/x.

By burning known mixtures of pure fat and carbohydrate in a

bomb-calorimeter (see Chapter 7) the number of kJ of heat generated, at a known non-protein RQ, for each litre of oxygen that is
used can be calculated. Tables or nomograms for the use of scientists and clinicians have been drawn up, and such a table is shown in
Table 1.1. It is seen, for example, that at a non-protein RQ of 0.7
each litre of oxygen consumed generates 19.62 kJ; and at a non-

Table 1.1. *Energy equivalent (in kJ) of one litre of oxygen used in oxidation
of mixtures of given weights (g) of carbohydrate and fat, at given non-protein
RQ values (Adapted from Lusk, 1928.)*

Non-protein Respiratory Quotient	Grams		kJ
	Carbohydrate	Fat	
0.70	0.00	0.51	19.62
0.71	0.02	0.50	19.64
0.72	0.06	0.48	19.69
0.73	0.10	0.47	19.74
0.74	0.13	0.45	19.79
0.75	0.17	0.43	19.84
0.76	0.21	0.42	19.89
0.77	0.25	0.40	19.95
0.78	0.29	0.38	20.00
0.79	0.33	0.37	20.05
0.80	0.38	0.35	20.10
0.81	0.42	0.33	20.15
0.82	0.46	0.32	20.20
0.83	0.50	0.30	20.26
0.84	0.54	0.28	20.31
0.85	0.58	0.27	20.36
0.86	0.62	0.25	20.41
0.87	0.67	0.23	20.46
0.88	0.71	0.22	20.51
0.89	0.74	0.20	20.56
0.90	0.79	0.18	20.62
0.91	0.84	0.16	20.67
0.92	0.88	0.15	20.72
0.93	0.92	0.13	20.77
0.94	0.97	0.11	20.82
0.95	1.01	0.09	20.87
0.96	1.05	0.07	20.93
0.97	1.10	0.05	20.98
0.98	1.14	0.04	21.03
0.99	1.19	0.02	21.08
1.00	1.23	0.00	21.13

protein RQ of 1, each litre generates 21.13 kJ. Since the litres of oxygen consumed in the 24-hour period have been measured and the non-protein RQ has been calculated, the kJ generated in the 24-hour period can readily be found.

A worked example is as follows:

A subject during a 24-h period consumed 400 l oxygen
$$\text{expired } 340 \text{ l carbon dioxide}$$
$$\text{excreted } \quad 12 \text{ g nitrogen}$$

$\therefore$ Protein metabolized $\dfrac{12}{0.16} = 75$ g

Oxygen used in protein metabolism $= 75 \times 966$ ml $= 72.4$ l

Carbon dioxide derived from protein metabolism
$= 75 \times 774$ ml $= 58$ l

$\therefore$ Oxygen used in carbohydrate and fat metabolism
$= (400 - 72.4) = 327.6$ l

and carbon dioxide used in carbohydrate and fat metabolism $(340 - 58) = 282$ l

$\therefore$ Non-protein RQ $= \dfrac{282}{327.6} = 0.86$

From Table 1.1, each litre of oxygen at this non-protein RQ generates 20.41 kJ

$\therefore$ 327.6 l oxygen produce $20.41 \times 327.6 = 6686$ kJ (from carbohydrate and fat)

and 72.4 l oxygen produce $18.57 \times 72.4 = 1344$ kJ (from protein)

$\therefore$ Total in 24 h $= 8030$ kJ.

When the method of indirect calorimetry is used for the experimental study of farm animals, the animal is placed for several days in a specially adapted farm-stall through which air circulates at a known rate, and the outflowing gas mixture is collected and analyzed for oxygen and carbon dioxide. Measurement of nitrogen excretion gives a measure of protein metabolism, and so the non-protein RQ can be calculated as described. For ruminants, the expired gas mixture contains methane, which is more energy-rich than carbon dioxide, and this must be taken into account. So to make an 'indirect calorimetry' method accurate requires at least as much apparatus as for 'direct calorimetry'. A very large part of our knowledge of the energy requirements of humans and livestock, of the changes at different periods of life and in different climates, of abnormalities in metabolism such as obesity and hormonal disturbance, has been obtained by these two methods.

A speedy though somewhat inaccurate method of indirect calorimetry is often used for clinical purposes. In this method, a non-protein RQ is assumed, not measured; the assumption is also made that the protein RQ is the same as this assumed non-protein RQ. The only measurement on the subject is the oxygen consumption in a known period of time, perhaps ten to thirty minutes. Oxygen uptake is measured from the rate of fall of the floating bell covering a spirometer containing oxygen-enriched air, with a carbon dioxide absorber. The spirometer is connected by a system of tubes to the subject's mouth, the tubes being supplied with valves to give a one-way flow. The process is quick and needs minimal equipment; and from the measured oxygen consumption, assuming a non-protein RQ of 0.82, the kJ per minute can be calculated, using a table such as Table 1.1. This method is used for detection of gross changes of metabolic rate, such as the differences between a normal and a thyrotoxic individual, and is thus a useful diagnostic tool.

BASAL, STANDARD AND RESTING METABOLIC RATE

The minimal metabolic rate observed when a human subject is at complete physical rest, emotionally undisturbed, in a post-absorptive state (at least 12 hours after the last meal) and in a room at a comfortable temperature, is termed his basal metabolic rate (BMR). This is the minimum metabolic rate associated with the activity of his tissues, including the work of his heart and respiratory muscles, and maintaining his body temperature. The terms in which this is defined are rather imprecise; 'comfortable temperature' for different people, or for the same person at different times of his life, may vary considerably, and the 'post-absorptive state' may be reached in only six hours if the last meal were a very small one, so that by 12 hours the person might be feeling the discomfort of hunger. More precision could be added by standardizing the conditions, to make them uniform between people, or between occasions: for example, making all observations at an ambient temperature of 20 °C, 12 hours after a meal containing 2000 kJ. This is called 'standard metabolic rate' (SMR). Further practical problems arise if the measurement is to be carried out on a species other than man, when one cannot rely on the subject's co-operation or make sure that it is 'emotionally undisturbed'. In this case the best one can do is to carry out the observations when the animal is showing all the signs of physical rest, e.g. body recumbent, limbs relaxed. This would be the 'resting metabolic rate'.

The effect of weight and surface area on metabolic rate, and the concept of metabolic body size as a basis of comparisons between species and ages, will be discussed in the chapter on heat production.

MEASUREMENT OF BODY TEMPERATURE

The surface temperature of a man or of other animals varies greatly with the ambient temperature, and in defining homeotherms as species which maintain a steady body temperature one refers to the deep body temperature or core temperature. But how can this temperature be measured? And indeed is it perfectly uniform throughout the core of an animal's long axis? Mercury-in-glass thermometers as used for clinical purposes can be placed in the mouth, the axilla or just within the anus, and will record the gross alterations of temperature of these regions which may be encountered in disease. The more precise measurements needed for scientific purposes have been much facilitated by the use of thermistors which can be made very small, and which can be calibrated with great accuracy. Such a thermistor with its associated wire can be swallowed by a human or other animal so that it lies at the bottom end of the oesophagus, close to the heart, or it can be inserted via the anus and pushed into the lumen of the rectum so that it senses the temperature of the abdominal cavity; it can also be placed against the tympanum of the ear, where it is very close to the arterial blood supply to the head. In experimental animals, such thermistors have been inserted directly into the hypothalamic region at the base of the brain, or alongside the cervical spinal cord, regions which have an important function in the control of body temperature. Core temperatures recorded in one animal from all these places simultaneously would not necessarily be identical. Indeed, for those species (such as some ruminants) which have a countercurrent heat exchanger in the blood vessels at the base of the brain, hypothalamic temperature may be lower than that of the core of the thoracic region. So it is important, in scientific study, to specify the actual site at which temperature is recorded, whether rectal, oesophageal, tympanic, or elsewhere.

Two further methods for measuring deep body temperature have recently been devised. For scientific study one can use a radio-pill: a miniature radio-transmitter calibrated to alter its transmission frequency with temperature. It can be swallowed by a human subject or an experimental animal, and it records the temperature of the stomach or regions of the intestine, until it is passed out from the rectum. For a general survey of the deep body temperature of a human population, the temperature of the freshly passed urine has been used. This can be obtained more conveniently than any thermistor measurement, and the urine temperature shows close agreement with the rectal temperature as measured simultaneously with a thermistor.

THE PLAN OF THE BOOK

The following three chapters describe heat production, and the many factors by which it is modified in normal life, and heat exchange with the environment by various routes and mechanisms. If production and loss balance perfectly, temperature remains constant: the following chapter describes the central mechanisms, assisted by peripheral sensors, which help in maintaining temperature stability, and the range and limits of this stability in different species and tissues. Next there follows a chapter on the particular problems experienced by the new-born and in old age in limiting fluctuations of deep body temperature. Then comes a chapter on adaptations to climate and acclimation, the process by which an individual moving to an unaccustomed thermal environment gradually accommodates to it. The total energy intake of an animal from its food, which may result in storage if not precisely balanced by energy output, is the subject of the next chapter. The way in which this energy storage modifies the body composition of humans and livestock, and the nutritional problem of obesity, are also described.

Finally there are two short chapters describing circumstances, one normal and one pathological, in which body temperature is raised. The chapter describing muscular exercise shows that the rise of body temperature may be an important factor in limiting athletic performance. The chapter on fever shows how the physiological processes maintaining deep body temperature continue to function but at a higher set point temperature.

CHAPTER TWO

Heat production

INTRODUCTION

A man sitting in a chair produces heat at the rate of about 80 watts. This is the rate at which heat arises from his metabolic processes, with about half coming from the gastro-intestinal tract, liver and muscle, and the remainder coming chiefly from heart, kidneys, nervous system and skin. If he has a meal, becomes active, or shivers in a cold room, his heat production increases; if he falls asleep it decreases. If he simply sits in the chair all the energy that is transferred in his metabolic processes appears as heat, so that his rate of heat production equals his metabolic rate. If he pulls on a rope and raises a weight at the other end of the room he does external work, and then his metabolic rate exceeds the rate at which heat appears in his body by an amount equal to the rate at which he is transferring energy to external work, including the heat dissipated by friction in the pulley system.

The rate at which heat is produced, however, is not necessarily the same as the rate at which it is released from the body as heat or external work. This occurs only in thermal equilibrium. The heat that is produced may be partially retained in the body, as at the beginning of muscular activity. Heat production then exceeds heat loss, and the mean body temperature rises. When the body cools in a cold environment, heat loss exceeds heat production and the *mean* body temperature falls, although the *deep* body temperature may be little changed. In thermal equilibrium, the rate of heat production is the same as the rate of heat loss. Also for many species there is a 24-hourly rhythm in the rate of heat production, activity and deep body temperature, so to find whether a human or other animal is in thermal equilibrium it would be necessary to make measurements over at least a 24-hour period.

FACTORS AFFECTING HEAT PRODUCTION

Apart from the effect of drugs, disease, emotional disturbance and activity, the chief factors that influence metabolic rate are body size, age, time of day, environmental temperature, food intake, and thermal insulation. Each of these factors will be discussed in turn. The units commonly used to express the rate of heat transfer (both production and loss) are watts, $kJ\ h^{-1}$ and (although now outmoded by SI units) $kcal\ h^{-1}$; 1 watt (W) = $3.6\ kJ\ h^{-1}$ = $0.86\ kcal\ h^{-1}$.

Body size

Under similar conditions, a larger type of animal generates more heat per hour than a smaller one, and an adult more than an infant. It might be supposed that body weight would form a satisfactory basis for comparison between large and small, as the heat is being generated throughout all the mass of actively metabolizing tissue; thus rate of heat production (or metabolic rate) might be expressed as watts (or kJ per hour) per gram or per kilogram of *body weight*. However, in homeotherms of temperate climates, rate of heat production is very closely related to rate of heat loss, and this in turn depends on the *surface area* (and thermal insulation) through which the heat is being dissipated. Larger animals have a smaller surface area, in relation to their volume or mass, than smaller ones. So it would be reasonable to express metabolic rate as watts per square metre of surface area. This is normally done in clinical studies, when metabolic rate has to be measured during diagnosis and treatment of hormonal or nutritional disorders. The surface area of a person cannot be conveniently measured directly, but is read off from a height–weight nomogram, or calculated from the expression: surface area in $m^2 = 0.1\ W^{0.67}$, where W = weight in kg.

Put more generally, metabolic rate, M, is related to body weight, W, by the expression $M = aW^b$, where a and b are constants. The term W^b is called *metabolic body size*. For the 'surface area' hypothesis, the exponent b has the value 0.67 or $\frac{2}{3}$. There has been much argument about the exact value of this exponent b, the ratio of the relative increase in metabolic rate to the relative increase in body weight. All authors agree that b has a value of less than 1, whether comparisons are being made within one species or between species of animals. Kleiber (1947) suggested that in comparisons between species, $W^{0.75}$ gave a better basis for metabolic body size than $W^{0.67}$. This is the basis shown on the 'mouse-to-elephant' double log plot in Fig. 2.1; and the $W^{0.75}$ value has also been used as a basis for calculating the food requirements of stock animals, as in Fig. 8.2. However, it has been calculated that because of random variations a nine-fold range of body size would be necessary to

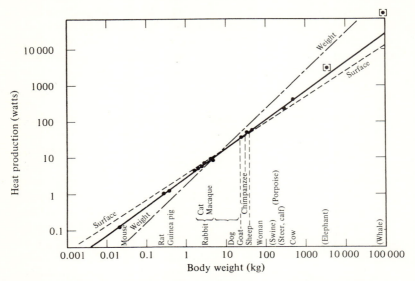

Fig. 2.1. Relation of the logarithm of metabolic rate and body weight in mammals. Also included are the slopes that would correspond with direct proportionality of metabolic rate to body weight, and to surface area. Redrawn from Kleiber (1947).

detect a statistically significant difference between $W^{0.67}$ and $W^{0.75}$ as a basis of comparison in a series of measurements of metabolic rates. Indeed, as Fig. 2.2. shows, the appropriate value of b can change in the course of the life of a single species, as in the case of man, where a higher value is appropriate between birth and puberty and a lower value later. The term $W^{0.67}$ is recommended as the metabolic body size for *intraspecific* use, unless there is some good reason for departing from this convention. The use of $W^{0.75}$ is mentioned here as the reader may meet the term in other connections.

Age

Young animals are of course smaller than adults of the same species, but age has an effect on metabolic rate which can be distinguished from the mere effect of body size. In all species, including man, metabolic rate increases rapidly in relation to body weight early in life, then more slowly later. In fact, in the new-born and young, the exponent b in the expression W^b for metabolic body size may be

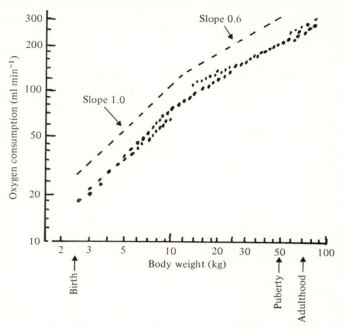

Fig. 2.2. Basal metabolic rate of man, as measured by oxygen consumption, related to body weight on double logarithmic coordinates, covering the entire period from birth to adult life. Data from various sources. Interrupted line: slopes of 1.0 and 0.6. Note that many of the results found during childhood lie on a slope approximating to 1, indicating a direct proportionality between oxygen consumption and body weight in this age range. Redrawn from Hill and Rahimtulla (1965).

close to 1: that is, metabolic rate is directly proportional to body weight in this age-group. For man, the initially fast and later slower increase in metabolic rate with weight is shown in Fig. 2.2. Following puberty, metabolic rate, per unit of surface area, tends to fall in both men and women; the rates at age 60 to 70 are lower than for young adults. This is shown in Fig. 2.3.

Time of day

Man is a diurnal mammal, active by day and resting at night. His metabolism is greatest and his body temperature is highest during the day, whereas the converse is true for a nocturnal animal such as the mouse. The 24-hourly variation in metabolic rate in another

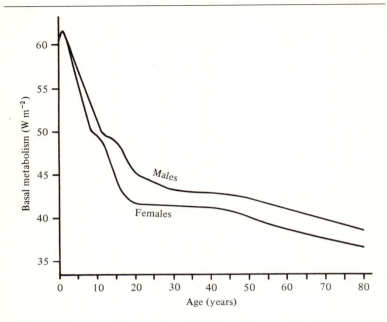

Fig. 2.3. Metabolic rate per unit surface area of humans (male and female) at various ages. Note that in both sexes there is a rapid fall to age about 20, a plateau between 20 and 40, and a slow fall thereafter. Redrawn from Guyton (1975).

diurnal mammal, the pig, is illustrated in Fig. 2.4, where it is evident that there is an amplitude of fluctuation of metabolism of about ±15% of the mean value over the day–night cycle. Short observations on metabolism, lasting only one or two hours, could therefore give misleading results if the effects of various factors were being investigated; differences that arise from measurements made at different times of day could be attributed to experimental or clinical treatments, whereas they may simply be due to variation within the 24-hourly cycle.

Since physical activity and ingestion of food have the effect of causing a large increase in metabolic rate, and since, for man, activity and meals are normally day-time occurrences, these circumstances would in any case make man's heat production higher in the day-time than at night. Continuous observations of the effect on metabolic rate of a series of common activities carried out over rather more than a 24-hour period are shown in Fig. 2.5. This sub-

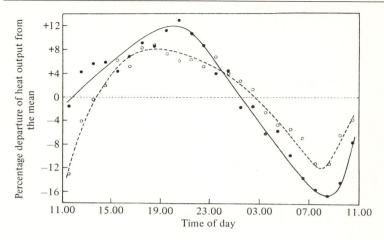

Fig. 2.4. Mean cycles of heat output, as percentage above or below mean level, for eight pigs subject to a regimen with 24-hour periodicity. Weight range 3–11 kg. ○–○ runs beginning at 11.00 h; ●–● runs beginning at 23.00 h. From Cairnie and Pullar (1959).

ject, a man of 48 years, with a body weight of 55.9 kg, had a metabolic rate, while sitting, of 90–100 W. The rate increased when a meal was taken, then to 110 W when the subject stood, and to 180 W when cycling with a low work load. During sleep overnight the metabolic rate fell to about 65 W.

Environmental temperature

The thermoneutral zone. It is characteristic of homeotherms that when they are exposed to low environmental temperatures they increase their metabolic rates, and, as a result, maintain their deep body temperatures. The inverse relation between resting heat production and environmental temperature is illustrated in section BC of Fig. 2.6 in which heat production is related to environmental temperature. When the temperature falls to B, the limit of heat-producing capability is reached, and if the surroundings become still colder, the deep body temperature begins to fall, metabolic heat production declines (the Q_{10} effect), and hypothermia and death occur if the environment does not become warmer. B is thus the cold limit on the environmental temperature scale.

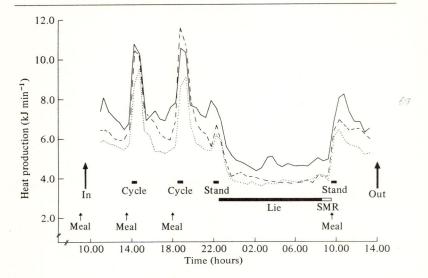

Fig. 2.5. Metabolic rate of a 48-year-old man weighing 55.9 kg during 28-hour periods in a whole body calorimeter, on three different occasions, with differing daily food intakes. Note the large fluctuations associated with meals and cycling, and the low level while lying down. From Dauncey (1980).

Conversely, with a rising temperature from B towards C the metabolic rate falls until it reaches a minimum value at C. C is termed the *critical temperature*; at environmental temperatures below this, metabolic rate must increase if the deep body temperature is to be maintained. Above the critical temperature there is a band of environmental temperature, CE, within which metabolic rate is at a minimum and is independent of environmental temperature: this is the *thermoneutral zone*. As the environmental temperature rises within the zone of thermal neutrality, temperature regulation is maintained by peripheral vasodilatation, relaxation of posture and sweating or panting, leading to increasing heat dissipation mainly by evaporation. At temperatures above E, however, the capability of the heat-dissipating systems is exceeded, the deep body temperature rises and the metabolic rate rises (Q_{10} effect again), with ensuing death in hyperthermia. E is the *hyperthermic point*, marking the hot limit; between the cold limit, B, and the hot limit, E, the deep body temperature is regulated by the increased heat production and

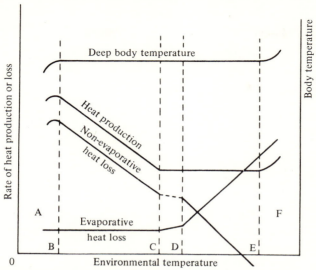

Fig. 2.6. Diagrammatic representation of relations between heat production, evaporation and non-evaporative heat loss and deep body temperature in a homeothermic animal. A, zone of hypothermia; B, temperature of summit metabolism and incipient hypothermia; C, critical temperature; D, temperature of marked increase in evaporative loss; E, temperature of incipient hyperthermal rise; F, zone of hyperthermia; C–D, zone of least thermoregulatory effort; C–E, zone of minimal metabolism; B–E, thermoregulatory range. From Mount (1979).

heat conservation (peripheral vasoconstriction, pilo-erection and compact posture) below the critical temperature, C, and by increased heat dissipation above C. The extent of the thermoneutral zone from C to E is sometimes questioned, because as E is approached sweating or panting increase considerably, which is not consistent with the concept of neutrality. The limited zone CD satisfies more closely the requirements of neutrality, with metabolic rate at a minimum but without a marked increase in the rate of evaporative heat loss.

The actual values of environmental temperature in Fig. 2.6 vary with species and with maturity. For example, the critical temperature, C, for an adult unclothed man is about 28 °C, and the cold limit, B, is about 14 °C, whereas for the new-born infant the corresponding values are 33 and 27 °C (Table 2.1). The arctic fox and

Table 2.1. *Critical temperatures* (°C) *and approximate cold and hot environmental limits for thermoregulation under still-air conditions, for the new-born
and mature unclothed man, pig and long-haired sheep*

	Critical temperature	Cold limit	Hot limit
Man			
new-born	33	27	37
mature	28	14	43
Pig			
new-born	34	0	36
mature	10	− 50	30
Sheep			
new-born	30	− 100	40
mature	− 20	− 200	40

From Mount, 1979.

eskimo dog probably have a critical temperature about − 40 °C. The
environmental temperature has so far in this discussion been taken as
the temperature of a standardized environment, in which air temperature represents the whole thermal environment, but in the non-stan-
dard surroundings found in practice (including, for example, the effects
of wind and sun) environmental temperature must be taken from a
composite of those factors that influence heat transfer between organism and environment. This topic is considered in the next chapter.

Shivering and non-shivering thermogenesis and brown fat. In Fig. 2.6
the range of environmental temperature from C down to B – which
for a man in normal indoor clothing might be about 15 °C down to
− 5 °C – is of great importance from the point of view of heat
production. Metabolic rate (in the resting post-absorptive
individual) is increased by two mechanisms: (1) shivering and (2)
non-shivering thermogenesis. Most mammals and some birds shiver
in the cold. At first, the slight cooling of the periphery and consequently of the blood supply to the hypothalamus has the effect, via
the motor nerves, of simply increasing the tone of skeletal muscle. If
the cooling is prolonged, there is actual shivering – small rhythmic
oscillatory contractions of the muscles – particularly of the jaws and
limbs. These two processes in skeletal muscle can double the rate of
heat production in the body as a whole.

Non-shivering thermogenesis occurs in many tissues of the body, particularly those rich in mitochondria and thus having a good supply of those enzymes associated with oxidative metabolism. At the cellular level, it consists of the uncoupling of oxidation from the synthesis of ATP and other energy-rich phosphate compounds, in consequence of which more oxidative cycles must be carried out to produce the amount of ATP required in the body; each oxidative cycle liberates energy as heat. One particular tissue in which much heat is produced by this process is brown fat, and this is a very important source of heat for the new-born mammal. (New-born humans are unable to shiver in the cold.) About two-thirds of the non-shivering thermogenesis can arise in brown fat in the new-born rabbit, for example, where the temperature at the site of heat production in brown adipose tissue may be 2–3 °C higher than colonic temperature.

Brown adipose tissue grows after birth at different rates in different species; it occurs characteristically in the inguinal, axillary and subscapular regions, and around the deep blood vessels of the neck. It can be calculated that 35 g of brown adipose tissue could use all the extra oxygen and produce all the extra heat associated with the human infant's response to cold exposure. Heat production in brown adipose tissue depends on a large supply of oxygen, and for this reason it is susceptible to hypoxia. Fat is the principal fuel, and the sympathetic nervous system appears to control thermogenesis in brown adipose tissue. Brown fat is important in the hibernating animal, where during the initial phase of awakening it may contribute 80% of the total heat production.

Brown fat is present in numerous vacuoles in the cell, as compared with the single vacuole of the white fat cell, and brown fat cells contain many mitochondria, indicative of the potentially high level of metabolic activity. When these cells are empty, they appear brown, although when full of fat droplets they are not nearly so easily distinguishable from white adipose tissue.

Reduction of heat loss. Numerous factors which alter the rate of heat loss from a man's or an animal's body, and thus the thermal demand of the environment, alter the rate of its heat production indirectly. Such factors as greater tissue insulation, a thickening of the layer of fur or feathers, shelter, and for humans clothing, lower the heat loss and thus indirectly the metabolic rate. So too do such behavioural responses as huddling together of litter-mate mammals or of nestling birds. These processes will be considered in Chapter 4.

Level of food intake

In the thermoneutral zone, metabolic rate is influenced by food intake. Here one must distinguish two quite separate effects, the one immediate, the other long-term. The immediate effect is associated with meal-times: metabolic rate may transiently double during and for about half an hour after the meal (Fig. 2.5). The long-term effect is related to the general level of food-intake. The mean metabolic rate of a man or other animal during a 24-hour fast is considerably lower than during a day of normal food intake but otherwise resting. Fasting metabolic rate for a growing pig is 380 kJ kg$^{-0.75}$ per day; if the pig is fed to maintenance level, it is 440 kJ kg$^{-0.75}$ per day. If the food intake is increased so that the pig is able to put on weight, the metabolic rate rises still more.

This 'heat increment of feeding' is of practical importance for two reasons. One is that it has the effect of lowering the man's or animal's critical temperature, the temperature at which a rise of metabolic rate is required for the maintenance of body temperature. This point is illustrated in Fig. 2.7 which shows heat production in a closely-clipped sheep, in two experiments: one when the sheep was on a high ration of food, and one on a low ration. On the low ration, the heat production in the thermoneutral zone was obviously much lower than on the high ration. But also, as ambient temperature fell, the sheep when on the low ration started increasing its metabolic rate at an ambient temperature of about 30 °C, but this did not occur until 20 °C when it was on the high ration. The heat increment of feeding has in fact taken the place of thermoregulatory heat production, in the temperature range 20 °C to 30 °C. It is likely that for all mammals, including man, diet-induced thermogenesis can partially replace cold-induced thermogenesis in this way. Above the critical temperature, the heat increment of feeding is simply added to the normal metabolism, and heat loss mechanisms eventually dissipate it. But the position concerning the critical temperature becomes important in the design of buildings for housing farmstock in the depth of winter in cold climates. The cost of artificial heating of buildings can be reduced if the animals are fed abundantly.

Another practical consequence is concerned with the growing animal, which is using its food supply not only for maintenance, that is, to cover its metabolic needs, but also to deposit energy as protein and fat. This deposition is never 100% efficient: its efficiency in thermoneutral conditions is in fact about 70%. This means that of an increment of energy intake (above the maintenance level), 70% is

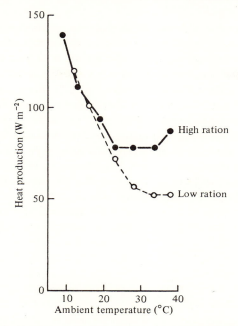

Fig. 2.7. Metabolic rate at various environmental temperatures in one closely clipped sheep during periods on a high ration (●) and on a low ration (○). From Alexander (1974), after Graham *et al.* (1959).

deposited as protein and fat, and 30% is dissipated as heat. It is important to consider this, when calculating the food requirements for humans and livestock. This point, and the different heat increments associated with high protein and high carbohydrate intake, are considered further in Chapter 8.

SUMMARY

The rate at which heat is produced (metabolic rate) in the body of a homeotherm depends on its size. In a large range of species differing widely in body size, it has been observed that metabolic rate is proportional to a certain function of their body weight, W, namely $W^{0.75}$, and this has been called metabolic body size. New-born and young growing animals have a higher metabolic rate, per unit of metabolic body size, than older animals of the same species. Man's metabolic rate declines rapidly until young adulthood, and thereafter more slowly. Heat production is greatly dependent on environmental temperature for all homeotherms. There is a range of environmental temperature at which metabolic rate is minimal, the low end of this range being called the critical temperature. Below this temperature, heat production increases by means of increased muscle tone, shivering, and non-shivering thermogenesis in which brown fat may play an important role. These processes maintain the animal's deep body temperature. Above the thermoneutral range, there occurs panting – or in humans sweating – and these mechanisms which promote evaporative heat loss also maintain deep body temperature, by dissipating heat. Raising the general level of food intake raises the metabolic rate; this is of some importance when livestock are housed at temperatures close to their critical temperature, because the critical temperature is lowered by abundant feeding.

CHAPTER THREE

Heat exchange with the environment

ROUTES OF HEAT EXCHANGE

Since most homeotherms have deep body temperatures in the range 36–42 °C, and since all the water surface and much of the land surface of the earth and the air close above it is below 36 °C for most of the time, it follows that the usual direction of heat flow is *from* the organism *into* the environment. However, circumstances and occasions occur in which an animal does gain heat directly from its environment, a reversal of the usual net direction of heat flow. This chapter has therefore been given the general title 'Heat exchange' rather than 'Heat loss', although the movement is in the direction of loss for most of the time.

The channels of heat exchange are radiation, conduction, convection and evaporation. Of these, evaporation and convection (the forced convection of blood flow and to a lesser extent the natural convection of air-flow over the skin surface) can be altered by physiological processes. The other mechanisms of heat flow (radiation and conduction) are only indirectly under physiological control through variations in skin temperatures, though they can be varied by behavioural responses. For example, cattle standing in the shade or desert animals dwelling in burrows during the day-time are reducing their intake of radiant heat; a dog stretched out on a cool stone floor on a warm day is increasing its heat loss by conduction. Heat is not necessarily transferred by all routes in the same direction, at any one moment: an animal or human standing in the sun on a windy day is gaining heat by radiation and losing heat by convection; a man sitting in a cool room may be simultaneously losing heat by radiation from his skin surface to the walls of the room but gaining heat by convection from a fan-heater. The illustration in Fig. 3.1, showing a gazelle resting in a hollow on a warm day, indicates possible routes of heat exchange with the environment. For assessment of the net direction of heat flow, and its measurement, the physical factors determining transfer of heat

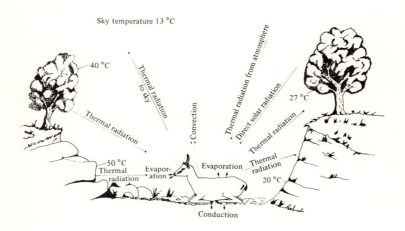

Fig. 3.1. Heat exchange of a mammal resting in a moderately warm environment. It is losing heat by conduction to the ground and also by natural convection and by evaporation; it is both gaining and losing heat by radiation. Note that the temperatures of trees and rocks on the sunny side of the hollow are above that of the animal's skin, which therefore receives radiant heat from them. Objects on the shady side, at temperatures below the animal's skin temperature, receive radiant heat from the animal. Redrawn from Gordon, Bartholomew, Grinnell, Jørgensen and White (1968).

listed in Table 3.1 must be considered. In this chapter, each heat transference mechanism will be discussed in turn.

RADIATION

Thermal radiation is of two distinct kinds, short-wave (or solar) and long-wave (or terrestrial). Short-wave radiation (wave-length 0.3–3 μm) comes from the sun and from extremely hot fires. Long-wave radiation (wave-length from about 5 to 100 μm) is emitted by all surfaces everywhere including surfaces of living organisms. So for short-wave radiation, organisms are always at the receiving end,

Table 3.1. *Factors that influence the different modes of heat transfer between organisms and the environment*

Mode of transfer	Animal characteristics	Environmental characteristics
Radiant	Mean radiant temperature of surface; effective radiating area; reflectivity and emissivity	Mean radiant temperature; solar radiation and reflectivity of surroundings
Convective	Surface temperature; effective convective area; radius of curvature and surface type	Air temperature; air velocity and direction
Conductive	Surface temperature; effective contact area	Floor or ground area temperature; thermal conductivity and thermal capacity of solid material
Evaporative	Surface temperature; percentage wetted area (for skin); site of evaporation relative to skin surface	Humidity; air velocity and direction

Adapted from Ingram and Mount (1975).

whereas for long-wave they may lose or gain heat, depending on whether the surrounding surfaces are cooler or warmer than themselves. In respect of heat balance, the appropriate measurement is that of radiant heat *exchange*: the algebraic sum of short- and long-wave radiation falling on the organism, and the long-wave radiant heat emitted from its surface. This depends on (a) the respective temperatures of the body and the surrounding surfaces, their areas and orientation to each other, and (b) the colour and reflectivity of the surfaces.

Area and orientation: short- and long-wave radiation

Since radiant heat travels in straight lines the exchange of heat between two radiating bodies depends on the area of the surface of each body which is 'seen' by the other, or the solid angle subtended by one body at the other. Camels orientate themselves when at rest during the day in a direction parallel to the sun's rays, thus minimizing the area of body surface exposed to solar radiation. Skiers are familiar with the sunburning of the underside of the chin and nos-

trils which catch the reflected solar radiation from the snow. The
same would apply to long-wave radiation: a person in a cold room
sitting close to a fire gains radiant heat, but only on the side of his
body facing the fire; the other side may be losing heat by radiation
to the cool walls of the room. The gazelle shown in Fig. 3.1 is
gaining radiant heat from sunny warm rocks, and losing it to a
shady cliff-face.

Colour and reflectivity

Short-wave radiation White surfaces reflect a fairly large proportion
of short-wave radiation falling on them, whereas black or dark
surfaces absorb most of this radiation. A commonplace example of
the effectiveness of a white surface to act as a solar heat reflector is
seen in the snow: sunburn can be caused by reflection of solar
radiation from snow surfaces even on overcast days, because 10–20%
of direct solar radiation can pass through thick cloud as diffuse
radiation. Again a person seated in the shade may receive short-wave
radiation by reflection from a nearby white wall. An antarctic ex-
plorer once suffered heat-stroke while lying on a sledge, as he gained
heat by solar radiation (direct or reflected from snow) more quickly
than he could lose heat through his heavy arctic clothing.

The 'black' skin of Negroes reflects only about 18% of the incident radiant heat from the sun, whereas the 'white' skin of Cauca-
sians reflects 30–40%. A practical consequence of this difference of
short-wave reflectivity was seen in an experimental comparison of
'white'-skinned and 'black'-skinned athletes. The two groups first
undertook training exercises indoors, and their rectal temperature
rose above resting level by about the same amount. They later performed similar training exercise out-of-doors in sunlight, and this
time the rectal temperature of the Negroes rose about 1 °C higher
than that of the Caucasian athletes. The additional heat load consequent on the greater radiant heat absorbance (lower reflectivity) of
the dark skin had caused a further rise of body temperature beyond
that due to the metabolic rate of exercise.

An experimental illustration of the response of black and white
surfaces to short-wave radiation is shown in Fig. 3.2, which depicts
a thermogram of the back of a man wearing a white shirt with a
black '3' marked on the back; in this thermogram, white shows the
highest temperature. When the man's back was exposed to a radiant

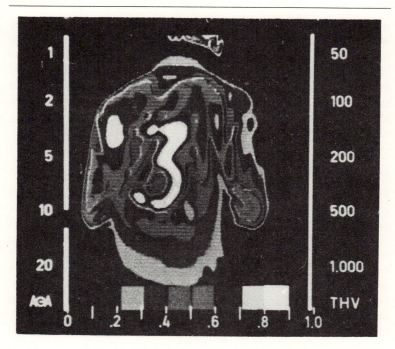

Fig. 3.2. Thermogram of a human subject wearing a white shirt with a black '3' on the back. There is a radiant heat load of 800 W m^{-2}, and the black '3' is 3 °C hotter than the rest of the shirt surface. The scale of shades on the abscissa progresses from colder on the left to warmer on the right. From Cena (1974).

heat source, the cloth on which the figure '3' was marked became 3 °C hotter than the rest of the shirt.

Long-wave radiation. The colour of the surface is of no importance in connection with the long-wave radiant heat exchange, which depends in this case on the temperature and emissivity of the surfaces; the higher the emissivity, the lower the reflectivity, and there are only very few surfaces that reflect appreciable amounts of long-wave radiation. Most surfaces absorb it, with the temperature of the object being raised and long-wave radiation then being emitted to surrounding cooler surfaces. This is sometimes called 'black-body emission' though it has no relation to the colour of the surfaces. Among the few materials that do in fact reflect long-wave in addition to short-wave radiant heat is smooth polished metal. A man sitting in a small experimental room lined with copper sheeting recorded the sensation of 'waves of heat' reaching him from the copper walls (this heat being a reflection of his own radiant heat output) although the walls felt cold to the touch. This ability of a metal surface to reflect body heat is the basis for the use of aluminium-foil 'space blankets' designed for astronauts and now widely used in survival kits.

CONDUCTION

Transference of heat by conduction depends on the temperature gradients between two bodies in contact, the area of contact between them, and the heat capacity and conductivity of their materials. Modern man in civilized societies rarely places a large area of his bare skin in contact with cool solid surfaces, so for most humans loss of heat by this route is trivial, though heat loss by conduction into *water* might be considerable. For many mammals, both domesticated and in the wild, a large heat exchange can occur by conduction. Sheep lying on cold ground outside in winter may lose 30% of their resting heat production by this route. Conversely, camels at rest on hot ground fold their legs below the body in such a way that the area of contact between skin and ground is minimal, and heat gain by conduction is reduced. Several domestic species such as dogs, cats and pigs use the amount of contact with ground or floor surface as a means of behavioural thermoregulation, and this may be a matter of importance in the choice of material for floor surfaces in animal housing. In some uncomfortably warm climates where the temperature scarely falls at night, people improve their comfort and ability to sleep by pre-cooling their bedsheets in the refrigerator, thus enhancing conductive heat loss when they lie down.

The amount of heat lost by conduction by warming cold food and drink to body temperature after ingestion is very small and rarely more than 3% of the total heat production.

CONVECTION

Convection means the transference of heat by the movement of a fluid (liquid or gaseous) medium from a warmer to a cooler region. It depends on the thermal capacity and flow rate of the medium, temperature difference, and, if the flow is over a surface, on the surface area. Convection includes the transference of heat from the core of an animal's body to the periphery by means of the blood circulation (forced convection or 'heat pumping'), as well as the transference of heat away from the skin surface by movement of air. So the first step of most of the heat loss by conduction or radiation from the skin, or by convective loss into moving air, would in fact be the convective transference of heat by the blood from core to the skin surface, thereby raising the surface temperature. This movement of heat by the blood is under physiological control, such control being one of the main processes by which heat loss can be regulated according to changing circumstances.

Transference of heat by air moving over the skin surface plays an important part in heat loss. Convection can be either natural or forced. An animal or unclothed human warms a layer of air adjacent to the skin. If the man is clothed, the outer surface of his clothing is warmed, though not up to skin temperature, by his body heat, and a layer of air outside this will be warmed. This warm air becomes less dense and rises, so a standing man would have a plume of warm air above his head. As it rises, the layer of warm air becomes thicker as the warm air around the legs flows upward over that already warmed around the trunk and head. This is illustrated in Fig. 3.3a where it is pointed out that by this means alone, 600 l min^{-1} may flow over the skin of a naked man.

For a clothed man, the heat loss by the route of natural convection is much reduced by the thermal insulation due to the trapping of a layer of still air close to his skin by his clothing; the fur and feathers of other animals would perform the same function. However, the upward streaming of warm air still occurs outside the clothing (Fig. 3.3b). One minor consequence of this natural convection is that the air which a standing or seated man breathes in, through his downward-facing nostrils, has been to some extent pre-warmed by its passage over the surface of his clothing, and the conductive heat loss from his respiratory surface is correspondingly reduced.

Forced convection by wind is an important consideration, especially where naked or nearly-naked surfaces are exposed, because the wind disturbs the unstirred layer of air adjacent to the skin, and so reduces the thermal insulation that the layer provides. At a wind speed of 1 m s^{-1}, the convective heat loss from a nude man is about 8 W per square metre of body surface, per °C temperature difference

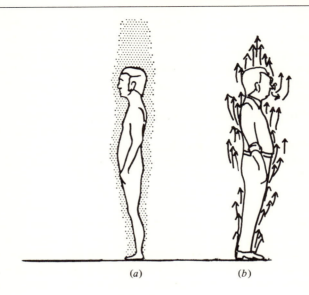

(a) (b)

Fig. 3.3. Natural convection over the surface of a standing man. The naked figure (a) illustrates the width (dotted area) of the convective boundary layer in still air. The direction of flow of air is shown by the arrows around the clothed figure (b): some of the air entering his nostrils would be partially warmed. For a naked man, the maximum flow velocity of air in natural convection is 0.5 m s^{-1}. The plume of warmed air may extend 1.5 m above his head and the volume of air moving upwards can be 600 l min^{-1} (a): from Clark and Toy (1975). By permission of *J. Physiol, Lond.* (b): redrawn from Edholm and Weiner (1981).

between skin and air; at a wind speed of 4 m s^{-1}, the convective loss is 18 W m^{-2} °C^{-1}. Further increases in velocity have diminishing effects in increasing convective heat loss. Obviously, it is important to expose minimal area of face and hands, in cold weather; conversely, forced convection by fans stirring the air over the skin surface is an effective way of cooling the body.

    The three routes of heat loss so far described – radiative, conductive and convective – are sometimes referred to collectively as 'sensible heat loss', in contrast with 'insensible heat loss' which means evaporation. This must not be confused with 'insensible *water* loss' described on p. 34. Confusion is avoided by the use of the terms 'evaporative' and 'non-evaporative' heat loss, the latter being the sum of heat loss by radiation, conduction and convection.

For all non-evaporative routes of heat flow, the units of measurements are watts per square metre of surface area per °C of temperature difference between surface and environment: $W\,m^{-2}$ $°C^{-1}$.

(i)　For conduction, the area would be the area of contact between the skin and the solid or liquid surface touching it.

(ii)　For radiant heat exchange, the area would be the mutually facing solid surfaces or zones. Radiant heat exchange, H_r (net) is given by

$$H_r = \sigma A(T_1{}^4 - T_2{}^4)$$

where A is the area of the mutually facing surfaces, T_1 and T_2 are the absolute temperatures in °K of the surfaces, and σ is the Stefan–Boltzmann constant $5.67 \times 10^{-8}\,W\,m^{-2}\,K^{-4}$. This is on the assumption that the surfaces are 'black bodies' absorbing all incident radiation and reflecting none, i.e. with emissivity 1. (For human skin of any colour, the emissivity for long-wave radiation is close to 1.)

For small differences between T_1 and T_2, a linear equation may be used as a close approximation: H_r (net) $= AK$ $(T_1 - T_2)$ where $K = \sigma 4T^3$, T being close to the mean of T_1 and T_2. For a radiant air temperature of 15–20°C and a skin temperature of 30–35°C, K is about $6\,W\,m^{-2}°C^{-1}$.

(iii)　For convective heat loss, it is difficult to write an equation covering all situations, since many variables are involved. For forced convection, an important determinant is wind speed, V. A decrease in the size of the body leads to greater convective heat loss per unit area, for a given skin-to-ambient temperature difference. So smaller animals, and parts of smaller radius such as fingers, are more susceptible to convective cooling. Other factors are the shape of the body, its orientation to the direction of flow, and whether the flow is laminar or turbulent.

(iv)　For evaporative heat loss from the skin and respiratory tract the units of measurement are watts per square metre of wettable surface per mbar vapour pressure difference between skin and ambient air. For man the wettable surface is considered to be 85% of the actual skin surface. Evaporative loss from the respiratory tract can be calculated from the volume of air respired per minute, by assuming that the expired air is saturated with water vapour at the temperature at which it leaves the body. (Latent heat of vaporization of water is 2500 J g^{-1} at $0\,°C$, 2400 J g^{-1} at $40\,°C$.)

EVAPORATION

The physics of evaporation

The driving force for the heat loss by evaporation is not (as it is for the three non-evaporative routes) a temperature gradient, but a gradient of absolute humidity, between that on the surface of the skin or respiratory tract and that of the ambient air. This is of great practical importance because it means that in conditions where the ambient air temperature approaches that of the animal's body, evaporation is the *only* effective route of heat dissipation. Absolute humidity is measured by the vapour pressure of water or by the mass of water vapour per unit volume. Relative humidity is the percentage of the saturation level of vapour pressure (or water-vapour holding capacity per unit volume) that is represented by the water vapour actually present. A thin layer of vapour adjacent to a wet surface (such as that of the respiratory tract, or of the skin in certain conditions) has a high absolute humidity and is normally 100% saturated. The ambient humidity in the air can be indicated by the wet- and dry-bulb temperatures (see below). The difference between the saturation vapour pressure and the ambient vapour pressure (which is sometimes termed the saturation deficit) is a measure of the drying power available.

A useful measure of ambient humidity is given by dry- and wet-bulb thermometers, a pair of thermometers measuring air temperature, one of which has a dry bulb and the other having a bulb kept permanently moist by a wick soaked in water. The wet-bulb thermometer records a lower temperature than that of the dry bulb, as a result of cooling by evaporation from the wick. The more vigorous the evaporation, the larger is the difference in the readings. The difference in readings can be translated into terms of relative humidity and water vapour pressure by means of the psychrometric chart (Fig. 3.4). As an example, a dry-bulb temperature of $25\,°C$ with a wet-bulb temperature of $12\,°C$ gives a vapour pressure of 5 mbar, and a relative humidity as low as 20%. Suppose now the air becomes wetter (say following a shower of rain without fall in tem-

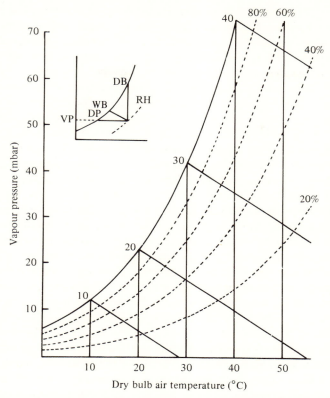

Fig. 3.4. Psychrometric chart relating dry (DB) and wet-bulb (WB) temperatures to vapour pressure (VP), dew point (DP) and relative humidity (RH). The dew point is the air temperature at which the given amount of water vapour in the atmosphere will start to condense. Other terms are defined in the text. Continuous curved line ——— temperature; interrupted lines - - - relative humidities. To find the relative humidity and vapour pressure for given wet-bulb and dry-bulb temperatures, find the point on the temperature line corresponding to the wet-bulb temperature, and draw a line from this point parallel to the sloping diagonals on the chart. The point at which this line crosses the vertical line corresponding with the given dry bulb air temperature marks the relative humidity, which can be read off on the dotted lines, or closer interpolated lines. From this point, a line drawn parallel to the abscissa will mark the appropriate vapour pressure on the ordinate scale and the dew point on the temperature line. (The inset on the chart shows this procedure.) From Ingram and Mount (1975).

perature). The dry-bulb temperature is still 25 °C, but the wet-bulb has now risen to 18 °C, as there is less evaporation from its bulb into the moister air. The vapour pressure read off from the chart is then 16 mbar, with 50% relative humidity. There would now be much less saturation deficit or drying power available for a human or other animal to lose heat by evaporation; the environment would feel noticeably less comfortable. The lower the ambient water vapour pressure, or relative humidity, the greater the gradient of absolute humidity between the skin or respiratory surface and the ambient air, and thus the greater the heat loss by evaporation.

When water evaporates, latent heat is taken up from the sur-roundings, the amount being 2500 J g^{-1} at 0 °C and 2400 J g^{-1} at 40 °C. When sweat evaporates from the skin surface, most of the heat is taken up from the skin, and only a little from the air, with consequent cooling of the body. When water evaporates on the surface of clothing or on an animal's coat, the clothing or coat and the surrounding air are cooled, but the cooling of the organism's body is much less effective. The site of evaporation is therefore important in determining the effectiveness of evaporative dissipa-tion of the heat from the animal.

Insensible water loss

Evaporative heat loss is biologically advantageous in that it can be regulated, and used as a means of dissipation of excess heat of the body in conditions so hot that there is little temperature gradient (or even an adverse gradient) for loss by any other route. However, there is inevitably some small continuous evaporative loss from all living organisms (and even from eggs and seeds) unrelated to re-quirements for heat dissipation. Since the skin of mammals and birds is not completely impermeable to water, and since their respi-ratory surfaces are always moist, there is necessarily a constant evaporation from the skin and respiratory tract of all terrestrial species, unless the air is 100% saturated with water vapour. The loss of water by these routes is called the *insensible water loss*.

A human at rest in a comfortable environment below 30 °C and of moderate relative humidity dissipates about 25% of his heat production in this way, and loses about 900 ml of water daily. For mammals and birds, the loss of both heat and water via the respira-tory tract is minimized by means of countercurrent heat exchange and water exchange in the upper respiratory tract. This countercur-rent mechanism works as follows: the continuous movement of air to and fro over the upper respiratory surfaces leads to evaporation in a manner similar to the effect of wind on a moist skin surface. The mucosal lining of the respiratory tract gives up water to the inspired air and so undergoes evaporative cooling. When the air is expired, some water is condensed on the cooled mucosal surface, which is below the dew point of air coming from the lungs, with a consequent release of latent heat. Some non-evaporative heat is also returned to the mucosa, and the air leaves at a temperature lower than deep body temperature. The heat and water exchanger is quite efficient, and by this means the temperature of air expired by birds is some-times closer to that of the environment than to that of the deep body; heat and therefore energy are conserved, and water is saved because the air that is eventually expired is not carrying as much water as it would if it were saturated at deep body temperature.

This process of water conservation via the respiratory tract is of particular value to some mammals living in deserts where the saving of water may be of even more importance than the dissipation of excessive body heat. The Kangaroo rat (*Dipodomys spectabilis*), for example, which lives without drinking water and derives its water only from metabolism and the slight moisture of food, has a very efficient respiratory countercurrent exchange, with expired air temperature as low as 24 °C; this minimizes water loss, but at the sacrifice of possible heat dissipation. The longer the respiratory tract, the greater is the opportunity for countercurrent exchange. A narrow tract, with large area-to-volume ratio is also advantageous.

In man at an ambient temperature of 20 °C, the warming of the respired air requires about 11 W, that is, rather more than 10% of the resting metabolism; at 0 °C, the corresponding heat loss is about 20% of the production. These values, although seemingly large, are smaller than would be the case if the countercurrent heat exchange did not take place. In humans in which the nasal passage is fairly short the tempera-ture of expired air in quiet breathing is 32 °C as compared with 37 °C at the alveolar surface. This temperature difference produces some con-servation of water and of heat. However, in extremely cold air, such as is met on Himalayan climbs, the loss of both heat and water via the respiratory passage is appreciable even in quiet breathing; climbers have to ensure that this lost water is replaced.

Panting and sweating

Different species dissipate excess heat by evaporation, either through the skin by sweating or through the upper respiratory tract by panting, or both. Man, other primates, horses and, to a lesser extent, cattle, are sweating species. Most carnivores, and also sheep, cattle, pigs and many birds, lose heat by panting. The whole wettable area of the naked skin of a human is large compared with the upper respiratory tract, so the skin route is to that extent advantageous. On the other hand, the loss of fluid from sweat glands is invariably associated with loss of some sodium chloride, whereas evaporation from the respiratory tract does not involve salt loss from the body unless the animal is also losing saliva, as in open-mouth panting. Furthermore, the activity of panting provides its own air movement, increasing the speed of evaporation; whereas sweating, in completely still air, might have the effect of creating a layer of saturated water vapour outside the skin which would soon diminish the effective evaporative rate. So each route of evaporative heat loss has advantages and drawbacks.

Panting consists of rapid shallow breathing which greatly increases the ventilation of the upper respiratory tract. In open-mouth panting the countercurrent mechanism of heat conservation may not operate, and the number of breaths per minute is so large that heat loss is considerable. In most species, the breaths during panting are so shallow that alveolar ventilation is not appreciably increased: excessive gas exchange at the alveolar surfaces, with consequent excessive loss of carbon dioxide leading to alkalosis, is thus avoided. In cattle, however, depth as well as rate of respiration may be increased during severe heat stress, and cattle may then indeed suffer from alkalosis.

Sweating as a mechanism of heat dissipation occurs in few species besides man. Sweating in horses can be profuse during exercise in response to adrenaline secretion or sympathetic stimulation. Thus it counters the accumulation of endogenous heat rather than exogenous heat. Man can sweat profusely in response both to the endogenous heat production of exercise and to high ambient temperature. He can thus survive in very hot but dry environments such as those of some deserts in the day-time, provided that he has an adequate water supply. The packing of the sweat glands of the human skin ranges from 2000 per cm^2 in the palm of the hand and sole of the foot, down to 100–200 per cm^2 in the limbs and trunk. The sweat secretion is more than 99% water; the concentration of sodium chloride, the chief solute, varies between 0.2 and 0.4 g per 100 ml. After frequent exposure to heat, the rate of sweating increases but the sodium chloride concentration tends to fall. The maximum sweating rates that have been measured in man are about 40 g per minute or rather more than 2 kg per hour. If all this evaporated on the skin surface, with no sweat dripping off in liquid form, it would provide an evaporative rate of heat loss of $(2400 \times 40)/60 = 1600$ W.

SUMMARY

The four channels of heat exchange between organism and environment are described, and the physical determinants of heat flow by these routes are discussed. Temperature gradient between the animal and the surroundings is a factor determining the direction and amount of heat flow by all non-evaporative routes. Panting and sweating serve many animals as a means of dissipating excess heat, and can be most effective in air of lowest relative humidity.

CHAPTER FOUR

Control of heat exchange

It was indicated in Chapter 2 that mammals and birds are able to increase heat production (metabolic rate) in response to a fall in ambient temperature, and this is obviously an important method of helping the animal to maintain a steady level of core temperature. It is not the only method, however. The control of heat losses to or gains from the environment by the routes described in Chapter 3 is an important aspect of the biology of all homeothermic species. The mechanisms of this control, whether physiological or behavioural, help the organism to maintain its deep body temperature within fairly narrow limits and thus make possible the continuance of its activity without the excessive fluctuations experienced by many terrestrial poikilotherms.

PHYSIOLOGICAL CONTROL

There are two main physiological controlling processes involved in heat exchange: skin vasomotor control, and evaporation by sweating or panting. The rate of flow of blood through arterioles and capillaries adjacent to the skin has a profound effect on the temperature of the skin surface, and therefore on the rate of heat loss by all three non-evaporative routes. Part of the range of variation is shown in the diagrams of Fig. 4.1, which illustrate the surface temperature of a man in cool conditions, when the temperature at the skin surface is only about 28 °C, and in warmer conditions, when the skin temperature is about 36 °C. In warm conditions heat loss is maximized by dilatation of skin arterioles allowing a high rate of flow of warm blood through the skin capillaries. In cold conditions the converse occurs: skin blood flow is minimized by vasoconstriction of arterioles, and heat is conserved within the deeper parts of the body. The gradient of temperature between skin surface and air, which is the driving force for non-evaporative heat loss, is thus reduced.

A variable countercurrent heat exchange between arterial and

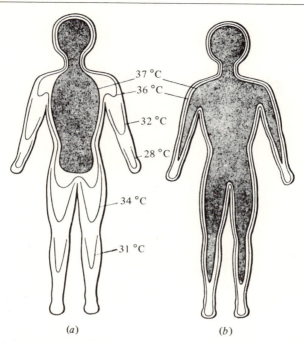

Fig. 4.1. The diminution of the size of the core in cool conditions by the redistribution of blood from the periphery to the core, shown by a series of lines connecting points of equal temperature. In (a) (cool environment) the 37 °C is within the head and trunk only. The temperature of the subcutaneous tissue of the hands and feet has fallen to 28 °C or lower. In (b) (warm environment) the 37 °C line is found close below the skin over a large part of the body surface. From Aschoff and Wever (1958).

venous blood in the limbs and their digital extremities enhances these effects. In cold conditions (Fig. 4.2a), a large proportion of the warm arterial blood flowing down the limb does not reach the skin surface and return from it by superficial veins, but instead passes through more deeply placed arterio-venous shunts, returning up the limbs in deep veins (venae comites) lying close to the arteries. By this close counterflow, there is a transfer of heat from the warm arterial blood to the cooler venous blood. By this arrangement not only is the blood returning to the central parts of the body not excessively cold, but also the extremities receive pre-cooled arterial blood; the temperature gradient across the skin surface and therefore the rate

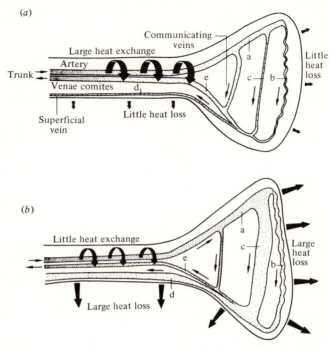

Fig. 4.2. Diagrammatic illustration of blood flow through an extremity (limb, snout, appendage) in (*a*) cold conditions; (*b*) warm conditions. (*a*): a, absolute flow through superficial tissues low, thermal conductivity equivalent to cork; b, flow within skin restricted to capillaries; c, little flow in arterio-venous anastomoses (shunt pathways); d, little flow in superficial veins; e, cold blood returns from the extremities in the venae comites; it is warmed by countercurrent heat exchange from the arterial blood. (*b*): a, absolute flow through superficial tissues high; b, some increases in capillary flow; c, large blood flow through arterio-venous anastomoses; d, most blood returns through superficial veins; e, little blood returns through venae comites. Intensity of shading indicates temperature of blood, width of vessels indicates relative blood flow. From Hardy (1979).

of heat flow through the skin to the environment is lowered. Indeed it has been observed that the temperature in the web between the claws of a gull's foot in winter may be almost as low as that of the ice on which the gull is walking (Fig. 4.3). In warm conditions (Fig. 4.2*b*), most of the warm arterial blood comes close to the skin surface and returns via the superficial veins, and heat loss is maxim-

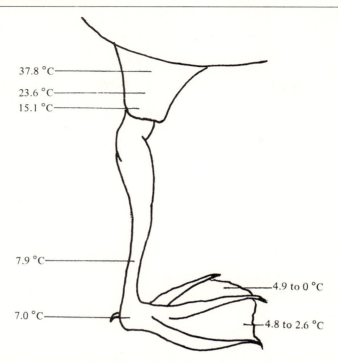

Fig. 4.3. Temperature in the leg of a gull standing on ice, in air at −16 °C. Redrawn from Irving (1964).

ized. The engorgement of the superficial veins of the limbs of humans in hot weather is a matter of common observation.

The variability of blood flow in skin capillaries of man is considerable. It has been calculated that even at ordinary comfortable skin temperatures the rate of flow in the skin capillaries is some ten times greater than the bare minimum required to supply oxygen and glucose to the cells of the skin. In very cold conditions the flow rate may be so much reduced that the skin metabolism starts to suffer, and it has been observed that the fingernails grow more slowly in arctic than in temperate or tropical regions. Conversely, in hot weather the amount of blood flowing to the skin may rise to 2.5 litres a minute, or half the cardiac output of a man at rest. It is not surprising that people with incipient heart disease may be more likely to reach the stage of heart failure in hot weather because of the extra load on the heart.

The mechanism of vasomotor control is the sympathetic nerve supply to the smooth muscle of the arteriole walls. In neutral conditions, impulses are continuously transmitted at a moderate rate down these nerves to the circular smooth muscles of the vessels, maintaining the diameter of the vessels at a moderate size. This is called vasomotor tone. An increase in the frequency of the impulses enhances the vasomotor tone and tends to shut down the arterioles by vigorous contraction of the smooth muscles. The lumen is reduced in size and blood flow lessened. A reduced frequency of nerve impulses, lowering vasomotor tone, allows the smooth muscle to relax, the lumen of the vessels to enlarge, and the blood flow to increase. The consequent warming of the skin provides heat loss by all routes; in addition the abundant blood flow provides adequate water in the sweating species for formation of sweat. The signals to the sympathetic nerves, enabling them to respond appropriately to external conditions, will be discussed in the chapter on temperature regulation.

The other main physiological control mechanism is the process of panting or of sweating which increases evaporative heat loss. Sweat is secreted onto the skin surface from sweat glands which are widely distributed in the human skin. Secretion of sweat is initiated and controlled by the activity of efferent cholinergic sympathetic nerves. The glands are capable of a sustained outflow of sweat, and the secreting process is eccrine, that is, involving transfer of fluid across the intact cell membrane. (In man the so-called apocrine sweat glands are restricted to the axillary and pubic regions and are not involved in thermal sweating.) During thermal sweating, the sweat glands also produce an enzyme, which acts on a plasma protein to produce the polypeptide bradykinin, which causes an intense, though short-lived, vasodilatation. It is not certain, however, whether the effect of bradykinin contributes significantly to skin blood flow in hot conditions. The control by the central nervous system of evaporative heat loss will be discussed in another chapter.

A physiological control mechanism seen in a few species, which reduces heat loss in a cold environment, is ptilo-erection (the fluffing-out of the feathers), shown especially by birds of the arctic; pilo-erection (or raising of hair) is the corresponding process shown by certain mammals (dogs, cats, horses, cattle) when moving from a warm to a cold ambient temperature. The process is controlled by sympathetic nerves in which an increase in the tonic impulses stimulates contraction of the smooth pilo-erection muscles at the base of

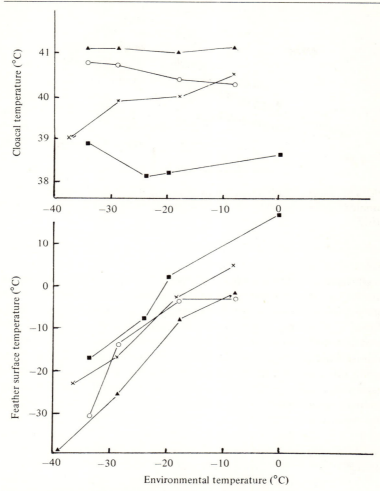

Fig. 4.4. Relation of feather surface and core temperature to environmental temperature in four species of birds. ■—■: black-capped chickadee; x—x: grey jay; ○—○: white-tailed ptarmigan; ▲—▲: raven. Redrawn from Veghte and Herreid (1965).

the hair or feathers, thus altering the orientation to the skin surface. The effect is to thicken the layer of trapped air outside the skin.

The effectiveness of this mechanism, at least for birds, has been assessed in a number of arctic species, in which ptilo-erection is often combined with a postural change (behavioural control). The

Fig. 4.5. Drawing from a photograph showing the huddled posture adopted by the Gray Jay at −35 °C. Redrawn from Veghte and Herreid (1965).

feathers are fluffed out and the bird adopts a hunched posture with the head tucked below the wing. The head is the warmest part of a bird's surface, as the blood flows close to the skin surface in this region, so the placing of the head below the wing protects this vulnerable part. The fluffed feathers trap a thick layer of still air, thus reducing convective heat loss; and the posture, which can make the bird's body assume an almost spherical shape, minimizes surface area. The success with which arctic birds can maintain their core (cloacal) temperature by such means is shown in Fig. 4.4, in which it is seen that although the temperature at the surface of the feathers falls nearly as low as that of the environment, down to −40 °C, the cloacal temperature hardly varies. Fig. 4.5 shows the spherical body shape, fluffed-out feathers and head-under-wing posture adopted by the Gray Jay, a subarctic species, at an ambient temperature of −35 °C.

The effectiveness of the corresponding process in certain mammals, pilo-erection, has not yet been fully studied. Observations on the calf have revealed that although the depth of the coat may be doubled by pilo-erection, its insulative properties are increased only 40%, not 100% as one might expect if the insulation were directly proportional to coat thickness. This may be because when the hairs are upright natural convection of air can occur to a greater extent in the spaces between the hairs, so heat is lost by convection as a proportion of the trapped air is not still. Pilo-erection in humans, present in the hair of head and limbs when one moves suddenly into the cold, has no physiological advantage.

BEHAVIOURAL CONTROL

Some of the processes involved in the behavioural control of heat exchange have already been indicated. In warm conditions man and other animals seek shade, minimizing radiant heat intake; animals lie on a cold surface, maximizing conductive heat loss. Many desert animals are nocturnal, remaining in burrows during the day and so avoiding intense radiant heat. In cold conditions, animals seek shelter from the wind, minimizing their convective heat loss. Many species of small mammals build nests or exacavate snow burrows for winter shelter. Man makes clothing and builds adequately insulated abodes which allow him to live in some degree of comfort in both the polar regions and the tropical deserts.

Daily torpor and seasonal hibernation, the states of low metabolism by which animals avoid the cold of the night and the food shortage of winter, might be considered as a combination of physiological and behavioural control; and perhaps also, in a sense, is the southward migratory flight of birds at the beginning of winter. Some of these processes will be considered in the chapter on body temperature.

Alteration of posture, one of the most obvious mechanisms of behavioural limitation of heat loss, is used by many species of birds and mammals, including man. Mention has been made of the placing of the bird's head below its wing, and many birds when roosting reduce heat loss from legs and feet by a seated posture which covers these limbs with the feathers of the body wall and wings; the domestic fowl is an example. A curled-up posture reduces the surface area through which heat can be lost by all non-evaporative routes. For a man standing upright, the effective radiating surface is in any case only 85% of his actual skin surface, since some surfaces (such as the upper arm and the chest-wall) would radiate to each other, and not to the environment. If the legs are bent at knee and thigh, the arms at the elbow, and the head is drawn forward and down, the surface for heat loss by radiation and natural convection can be much reduced.

A striking example of effectiveness of reducing surface area available for heat loss is seen in young animals, particularly pigs and rodents, born as a litter, with very little hair covering. Such litters huddle together so that the effective surface area of six piglets would be less than six times that of one piglet, and the heat losses by radiation and natural convection would be reduced correspond-

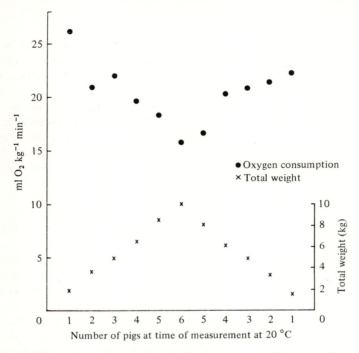

Fig. 4.6. The relation between metabolic rate per unit of body weight and the number of new-born pigs together in a chamber at an environmental temperature of 20 °C. From Mount (1960).

ingly. This is illustrated in Fig. 4.6. A single piglet's metabolic rate (indicated by oxygen intake) was measured in a temperature-controlled room at 20 °C, a temperature below the piglet's critical temperature so that its metabolic rate gives a measure of the thermal demand of the environment. Another piglet was then placed in the pen, and a further measurement was made; then another piglet, and another, until a group of six piglets was present. The piglets were then removed one by one, and measurements of oxygen consumption, as an indication of heat production and thus of heat loss, were made after each addition and withdrawal of a piglet to and from the group. The figure shows the rate of oxygen consumption, per unit of piglet body weight, at each stage, as the group increased in number and then decreased. The single piglet consumed about 25 ml oxygen kg^{-1} min^{-1}, whereas the group of six, huddled together, consumed only 15 ml oxygen kg^{-1} min^{-1}; this is approximately 60% of the oxygen required by the six animals if they had been housed separately and not allowed to huddle together, if all the piglets maintained similar deep body temperatures. This illustrates the success with which huddling can reduce the thermal demand of the environment.

In a warm environment, in addition to the spread-out posture commonly adopted by many mammals, certain species show a type of behaviour which promotes evaporative heat loss from the skin. Pigs and hippopotamuses wallow in shallow mud or water, and pigs wallow in their own urine if no other moisture is available. Some small mammals lick their fur, spreading saliva which cools the skin as it evaporates from the surface of the fur. The effectiveness of this behaviour is limited by the availability of drinking water to the animal; if this is absent or greatly reduced the animal cannot salivate sufficiently.

THERMAL INSULATION

Thermal insulation can be defined as the temperature difference across a medium of unit surface area which can be maintained when unit heat flow is taking place through it: its units are $°C$ m^2 W^{-1}. The reciprocal of this, thermal conductance, is the heat flow per unit temperature difference: W m^{-2} $°C^{-1}$.

Thermal insulation provides the resistance to the flow of heat from the core of the animal's body to the outside air. It consists of three insulations in series: the tissue insulation, I_t; the resistance to the flow of heat provided by fur, hair, clothing or feathers at the skin surface, I_{cl}; and the resistance to heat transfer from the outer surface of clothing or coat to the surroundings, I_a. Total insulation, I, per unit area is the sum of these quantities, $I = I_t + I_{cl} + I_a$.

$$I = \frac{\text{Temperature difference}}{\text{Heat flow}}$$ and has the units $°C$ m^2 W^{-1}.

$(I_{cl} + I_a)$ is called the 'external insulation'. I_a has the value of about 0.11 $°C$ m^2 W^{-1} in still air.

It is of some physiological and practical interest to measure I, or I_t, or I_{cl}. For instance, various clothing materials have different values of I_{cl}; similarly the I_{cl} of an animal's summer and winter pelt may be compared; comparison can also be made of the I_t of a fat and a thin animal, or of an animal during peripheral vaso-constriction and peripheral vasodilatation. Although for clothing a value for thermal insulation can be obtained by a direct test-measurement, the insulation of the tissue, or of an animal's pelt in life, has to be obtained by calculation from heat flow and temperature difference, $I_t = (T_b - T_s)/H$ where T_b and T_s are the deep body and skin temperatures, and H is the total heat generated in (or lost from) the body in unit time.

At the surface, the total heat, H, is dissipated by non-evaporative (H_n) and evaporative (H_e) routes. The external insulation is concerned only with H_n and this value can be obtained either directly (by placing the person or animal in a calorimeter), or indirectly as $(H - H_e)$. The external insulation, $(I_{cl} + I_a)$, is $(T_s - T_a)/H_n$, where T_s and T_a are skin and air temperatures respectively.

The 'clo', used as a measure of the insulative effect of clothing or fur or feathers and defined below, has the numerical value $0.155\ °C\ m^2\ W^{-1}$. It is the insulation of normal indoor clothing of a sedentary person in comfortable surroundings.

Tissue insulation

This constitutes the body's internal insulation and is the barrier to conductive heat flow provided by the body tissues. It has been calculated to be about the same as that of cork for the superficial tissue. The thicker the layer of subcutaneous fat, the greater is the tissue insulation. It must be remembered, though, that most of the movement of heat from core to skin is carried out not by conduction through tissues, but by the forced convective movement of blood, the subcutaneous layer of blood vessels being outside the subcutaneous fat layer. So although the fat may help to prevent heat loss in cold conditions in the animal or man at rest, when the cutaneous blood flow is not very great, it is a much less effective impediment to heat dissipation in hot conditions or in exercise, when cutaneous blood flow is enormous.

In a severe heat wave in the United States during the 1930s when there was a particularly high incidence of death from heat stroke, it was observed that many of the victims were unusually fat, so the question of fat as a barrier to heat dissipation was considered. However, it was also noted that many of the victims had a history of heart disorders, conditions often associated with excess body weight. The heat stroke deaths were more probably related to the strain of the large cutaneous blood flow on an already strained heart, than to the fat layer as a barrier to heat loss.

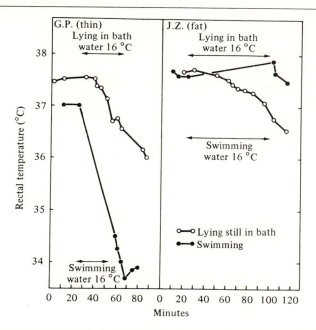

Fig. 4.7. Rectal temperature of subjects G.P. (thin) and J.Z. (fat) in water at 16 °C, with and without swimming. From Pugh and Edholm (1955).

In cold conditions, on the other hand, man's subcutaneous fat layer may be of great importance, even life-saving in cold water. This is illustrated in Fig. 4.7, which shows several interesting points. One is that whereas heat loss by conduction is trivial for a man on dry land, it is quite considerable for a man in water, the water at 16 °C touching every part of his skin surface. Even the fat man in Fig. 4.7, lying still and thus minimizing convective heat removal from his skin, shows a drop of rectal temperature of about 1 °C in the 2-hour period; the temperature drop for the thin man is about 1.5 °C in 90 minutes. During swimming, the thin man's rectal temperature shows a rapid fall. The additional rate of blood flow through the actively working muscles means more rapid removal of heat from core to periphery, and the effective thermal insulation of his body tissue decreases. In the case of the fat man, making the same standardized swimming movements for a longer time, although the flow-rate in his skin blood vessels must also have increased, the additional thermal insulation provided by his fat offsets

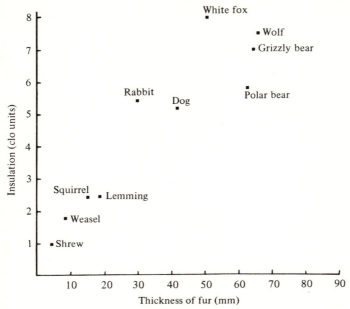

Fig. 4.8. Insulative capacity of fur from various mammals. Redrawn from Scholander, Walters, Hock and Irving (1950).

this effect and his rectal temperature is maintained. Further observations made in water at various temperatures have shown that in water at 25 °C, exercise has little effect on the rate of heat loss or fall in rectal temperature; in water at 35 °C, exercise converts a fall in rectal temperature to a rise.

External insulation

This comprises the clothing or animal coat and the surface insulations. A convenient though arbitrarily defined unit for measurement of clothing or coat insulation is the 'clo' unit, already mentioned above. It is defined in general terms as the insulation provided by the clothing worn in a comfortable room temperature, and more precisely as the insulation required to maintain a temperature difference of 0.18 °C, with a heat flow of 1.16 W m^{-2}. For many mammals it has been found that the insulating capacity is roughly proportional to the thickness of the coat or pelt. This is illustrated in Fig. 4.8, in which the insulating capacity in clo units is plotted against thickness of the fur in mm. This proportionality however is, for mammals in general, by no means precise: a sheep's fleece, for example, 100 mm thick gives insulation that is not 20 times that of a 5 mm fleece, but only about 8 times. Part of the explanation for this may be that as the thickness of the fleece increases so also does the surface area from which heat is lost. The texture and the closeness of the fur, wool or hair, as well as its thickness, is also important, thick straight hair giving less insulation than long loose fibres. The thickness of the layer of fur is limited by the requirement for movement: the smaller species of mammals necessarily have less fur thickness than larger ones.

The insulation of animal's hair provides a thermal barrier in both directions. This was shown by observations on the camel of which the thickness of fur ranges in length from 30 mm to 140 mm across various parts of the body. Such a camel in certain standard conditions lost 2 l water per 100 kg body weight per day, during heat dissipation by evaporation. When the camel was shorn, to give a fur thickness of less than 10 mm, an additional heat load entered from the environment and the camel's core temperature could be maintained only by a greater evaporative loss, which now became 3 l per 100 kg per day. The thick layer of wool of the merino sheep in the hotter parts of Australia provides a similar thermal barrier to the entry of radiant heat.

Clothing

The insulative properties of various types of clothing have been studied in some detail, and requirements for the comfort of polar and high altitude explorers, astronauts and skin-divers have been a great stimulus for such study. On earth the essential requirement for protection against cold is the trapping of a layer of still air outside the skin, in the interstices of the cloth; in general the thicker this layer of air the better the insulation, as was seen with the fur thickness of various mammals. One clo keeps a person in thermal comfort at a temperature of 22 °C (a warm room) with relative humidity less than 50% and an air movement of 0.1 m s^{-1}. Two clo units keep the person comfortable at 15 °C, three clo units at 8 °C, and so on, each additional clo allowing a drop of air temperature of about 7 °C. But the bulkiness of clothing is so great at 6 clo that further increases are impossible. Astronauts are protected in the intense cold of outer space partly by clothing in which hot water circulates in tubing between the layers, and partly by clothing which reflects internally a large proportion of their own body heat, which is thus never lost to the environment. In water, the essential requirement is that cold water should not touch the greater part of the skin surface, so that heat loss by conduction to the water is avoided. The heavy rubber suits of skin-divers provide good insulation through the rubber. Water leaks in through gaps at wrists or ankles and is trapped inside the suit and is soon warmed to body temperature. This, together with the diver's sweat, provides a layer of warm moisture immediately adjacent to his skin. However, the thermal conductivity of the rubber, though low, is not negligible, so the danger of cooling limits the safe period of skin-diving in cold water.

The main problem with outdoor clothing is the necessity for protecting it from rain and at the same time allowing vaporization of sweat and insensible water loss from the skin. If the clothing becomes soaked so that the air trapped in the fabric is replaced by water, its insulative property becomes scarcely greater than that of the cotton, wool or other fibres of which it is made. On the other hand a completely waterproof outer layer traps the body's moisture close to the skin, and in polar regions, this can be highly dangerous as the moisture may then freeze solid. Certain very finely-woven fabrics cause rain water to run off instead of soaking in, and still permit evaporation of the body's moisture through the interstices. Silicone-spraying of the outer surface of fabrics has a similar effect. Leather and bark cloth, moderately waterproof, may trap a layer of air between the coat and the skin, but unlike woven or knitted fabrics have no interstices in which to trap still air.

In warm conditions the problem of clothing is very different. White or light-coloured clothing reflects a high proportion of heat (solar radiation) but allows some heat to penetrate to the body. Dark clothing absorbs most of the incident heat, which is then available for removal from the outer surface of the clothing by convection caused by wind. An important requirement of tropical clothing is that it should be loose, so that the body's movements will cause the clothing to move about, creating draughts close to the skin and so promoting heat loss by convection and evaporation. Cotton and thin wool are more suitable for tropical clothing than is nylon. The fibres of cotton and wool readily absorb the moisture of the sweat, which is then re-vaporized from the outer surface of the cloth. So water vapour is removed down a vapour pressure gradient both through the interstices and from the fibres themselves. Nylon does not absorb moisture so readily; a layer of air saturated with vapour is formed between the cloth and the skin, and although water vapour is still being slowly removed through the interstices of the nylon, this removal may not keep pace with the sweating rate. The sweat cannot evaporate into the already saturated layer of air close to the skin, and rolls off the body, contributing nothing to its cooling. In the desert, clothing gives protection, both against the heavy radiant heat load of day-time, and the cold, sometimes below 0 °C, at night. The heavy but loose floppy clothing worn by experienced desert-dwellers such as the Bedouin is very satisfactory for its purpose.

SUMMARY

One of the major mechanisms by which body temperature is controlled is the re-distribution of blood, from centre to periphery in warm conditions, or withdrawal from the periphery to more central regions in the cold. This process by altering skin temperature will alter the heat available for removal by the three non-evaporative routes. In addition heat can be lost by evaporation; this loss can be greatly increased by physiological means and this increase forms another major mechanism for control of body temperature. Behavioural control, including alteration of posture and huddling, is a further means of altering heat exchange. Insulation outside the skin is provided by the formation of a layer of still air trapped in the interstices of fur, feathers or (for man) clothing.

CHAPTER FIVE

Body temperature and its regulation

Over a period of time, there is, in the bodies of homeotherms, a balance of the heat generated in the body and gained from the environment (as described in Chapters 2 and 3) against the heat lost from the body, as described in Chapter 3. The result of this balance is that the deep body temperature remains stable within certain limits. The first part of this chapter will attempt to answer a number of questions concerning body temperature. What is the actual value of the deep body temperature of various species of homeotherms, and within what limits does it normally fluctuate? What is the range of variation in different tissues within the body? What are the limits of variation compatible with life? The next part of the chapter describes what is known in general of the central control mechanisms by which a certain degree of stability of deep body temperature is contrived, and the routes by which information about the heat generated, gained and lost reaches these central control mechanisms, thus enabling them to function. Finally a short section is devoted to a description of controlled hypothermia of humans, which enables certain surgical procedures to be carried out.

BODY TEMPERATURE OF MAMMALS AND BIRDS

Various values obtained for the deep body temperature, usually measured in the rectum or base of the oesophagus in mammals and the cloaca in birds, are shown diagrammatically in Fig. 5.1. A single value, or even a range of values, for deep body temperature must be treated with caution. Environmental temperature, exercise, stress (including the stress of handling for making the measurement), food intake and time of day might cause considerable variation, particularly for birds and small mammals. For some of the species listed in Fig. 5.1, only two or three individuals have ever been studied; and the 'range' signifies, in some cases, the range of variation in one individual over a 24-hour period, while in other cases, the variation among a group of individuals of the species.

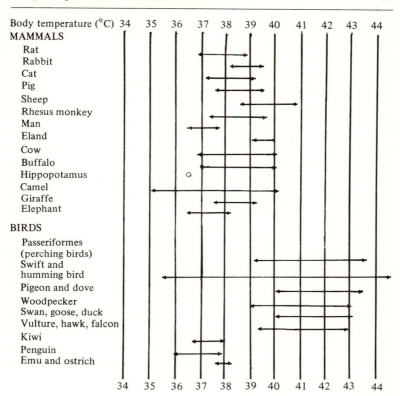

Fig. 5.1. Deep body temperature of various species of mammals and birds. The long lines for camel and for swift and humming-bird indicate the large 24-hourly fluctuations of body temperature of these species. Data from Whittow (1971), and from Bligh and Harthoorn (1965).

The numerous values obtained for domestic species accustomed to handling, such as pig, sheep and cow, are likely to be more reliable than those few obtained for, say, wild rat or buffalo. In general it seems that (a) values for the larger domestic animals – cow and sheep – are above that of man; (b) of the larger mammals, the camel and the buffalo show a considerable 24-hourly variation of about 3 °C, and the sheep, like man, a much smaller variation, of about 1 °C; and (c) most species of birds when awake but at rest have temperatures in the range of 40–42 °C, and nearly all birds show a marked 24-hourly variation, depending on their activity and food intake. A 24-hourly cycle of temperature fluctuation has been observed in all species so far studied and parallels the fluctuation of metabolic rate mentioned in Chapter 2. The extent of the fluctuation varies between species and with conditions such as exercise. The fluctuation tolerated by the camel, which can be as great as 6 °C in conditions of dehydration, makes it well able to withstand desert conditions: heat is stored in the body in the day-time and lost at night when ambient temperature drops.

There has been some speculation about the slightly higher temperatures of birds compared with mammals, and one suggestion is that of a connection with the high energy demand of flight. This necessitates a small body weight: 95% of all species of birds weigh between 2 and 1000 g. This in turn implies a high surface-to-volume ratio, and a rather thin insulating layer outside the skin. Thus, in the face of an inevitably rapid heat loss, a very vigorous metabolism in the heat-generating deep tissues would be required to attain a reasonable mean temperature throughout the body, so the core temperature would be high. Many species of mammals, on the other hand, are in the range of body weight 5 to 500 kg, which provides more scope for insulating capacity: a somewhat lower rate of metabolism in the core would thus be adequate to maintain a suitable mean temperature. The idea of an inverse relation of body size and body temperature cannot be pushed too far, however. There are many small birds with body temperatures below 40 °C; and among the mammals, the cow has a body temperature of 39 °C similar to that of the much smaller squirrel, and both are above that of a species of intermediate size, man.

TORPOR AND HIBERNATION: 'ADAPTATIVE HYPOTHERMIA'

In some species of birds and mammals – all of them species of small body weight – the body temperature and metabolic rate fall greatly during the inactive period of the 24-hour cycle, a marked exaggeration of the 24-hour cyclical rhythm common to homeotherms. This state of low temperature and metabolism is called torpor: by it the animal is able to avoid temporarily the necessity for continual food intake required by the high metabolic rate of its active phase. Insectivorous bats are in torpor during the day-time, emerging at dusk and feeding actively during the night. Humming-birds are in torpor at night, and feed during the day; many species use highly calorific flower nectar. During the torpid state, body temperature may fall by 8 or 9 °C.

A far greater fall in body temperature is experienced by those mammals which undergo hibernation during the winter. Some hibernating species have a deep body temperature of 10 °C or lower. The metabolic rate is extremely low: the respiratory rate may be reduced to once per minute and the heartbeat to 5 or 6 times a minute. Hibernators usually seek a hole, burrow or cavity which provides some shelter from the severity of the winter and gives the animal a fairly stable microclimate. All hibernators periodically arouse from their winter sleep – generally spontaneously at fairly regular intervals. They become increasing arousable however as the time for spring arousal approaches; this also occurs when there is a drop in the ambient temperature to a level which threatens to cause body temperature to fall below the limit of tolerance. During an arousal period, the body temperature increases rapidly and the animal may excrete, and eat some stored food, before returning to the hibernating state. At the time of arousal, whether during or at the end of winter, the increase in body tissue temperature occurs first in the heart, lungs and brain, and later in the abdominal organs and extremities.

The heat is supplied partly by shivering but largely by the metabolism of brown fat. A finely adjusted vasomotor control distributes the blood flow and thus the heat to the regions of the body in sequence. Fig. 5.2 shows the temperature at the head (cheek-pouch) and rectum of a hibernator, the Golden Hamster (*Mesocricetus aureatus*), during its arousal from hibernation. It can be seen that the rise in temperature in the rectum lags behind that of the cheek-pouch, so that for a time there is nearly 20 °C difference in temperature between the two ends of the animal's body.

The low body temperatures experienced in the torpid and hibernating states are clearly compatible with life, and these states of adaptative hypothermia are ways in which certain birds and mammals avoid the vicissitudes of temporary cold and food shortage. The homeotherm during hibernation may even become temporarily poikilothermic, showing a deep body temperature only a few degrees above that of the ambient air.

TEMPERATURE OF TISSUES AND ORGANS

The hottest parts of the homeotherm's body are the heart, the liver and (in certain conditions) the brown fat – the places from which a large part of the resting body heat is generated and distributed. The temperature of other organs for the animal at rest depends greatly

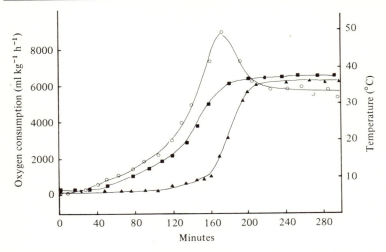

Fig. 5.2. Body temperature during arousal from hibernation in the hamster (*Mesocricetus aureatus*). During hibernation, temperature of rectum and cheek-pouch are about 10 °C. At arousal, the rise of rectal temperature (▲–▲) lags behind that of the cheek-pouch (■–■). Oxygen consumption (metabolic rate) (○–○) rises quickly and overshoots the fully-awake level. Redrawn from Lyman (1948).

on the blood flow from moment to moment. Certain particular organs and tissues which have their own specific temperature requirements, or which take part in temperature control of the whole body, will now be considered.

Skin

As has been indicated in Chapter 4, and Figs. 3.1 and 4.2, the skin is the tissue through which nearly all non-evaporative and some evaporative heat is lost to the surroundings, and skin temperature determines the rate of non-evaporative heat loss. A cool shell of skin and body wall, as provided by a reduced blood flow consequent on increased vasomotor tone, protects the warmer thoracic and abdominal organs; in some extreme conditions the surface of the skin may temporarily be as cold as the ambient air. The countercurrent system of blood vessels of the skin (Fig. 4.2) ensures that even during this intense vasoconstriction a small amount of (pre-cooled) blood reaches the skin surface. In man, in severe or prolonged cold, the blood vessels of the skin may relax again, since cold makes the

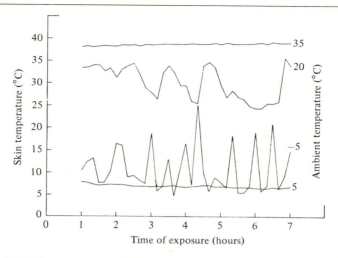

Fig. 5.3. Skin temperature of the left ear of a calf at various environmental temperatures, as shown on the right-hand side of the graph. From Whittow (1962).

smooth muscle of the blood vessel walls insensitive to vasoconstrictor agents. Blood may then flow into the passively dilated skin capillaries, and areas of skin which had been white become red; blotchy patches of red skin on the faces of people on a cold winter day are a common sight.

If this cold vasodilatation occurs all over the body, as might happen during immersion in cold water, heat is rapidly lost, the core temperature falls, and intense shivering starts. However, cold vasodilatation limited to certain parts of the body such as the fingers may have a biological advantage in preventing frostbite. An intensely vasoconstricted finger is being deprived of glucose and oxygen and this lack might eventually lead to irreversible tissue damage. If, however, cold vasodilatation occurs, and at the same time the muscles of the fingers and the hand are rapidly moved, producing a brisk flow of warm blood from the deeper parts of the body to the extremities, the danger of frostbite is averted, and the skin of the fingers becomes warm and pink. Many people have seen and marvelled at the hands of fishmongers who are constantly handling and cutting ice-cold fish, apparently without discomfort – an illustration of the beneficial effect of local cold vasodilatation.

There have been a number of observations of cold vasodilatation in species other than man. Fig. 5.3 records the skin temperature on the ear of a calf during a 7-hour exposure at each of four different ambient temperatures. The skin temperature remained fairly steady at 35 °C and at 5 °C. It fluctuated at 20 °C, and there were more marked and rapid fluctuations at − 5 °C. At this temperature, the bouts of vasodilatation transiently raised the skin temperature into the range 15–25 °C, its temperature falling back to about 6 °C between these phases of vasodilatation.

Testis

The testes of the adult male of many mammalian species are held in a fold of connective tissue, the scrotal sac, outside the abdomen and thus at a slightly lower temperature than the other abdominal organs. The scrotal sac maintains a temperature between 2 °C and 7 °C lower than the core temperature. Even in the seal, in which the testis lies not in a scrotal sac but under the blubber near the animal's hindquarters, the testis maintains a temperature cooler than the core. This may be because cool venous blood returning from the hind-limbs passes through the blubber close above the testes.

It can be shown that the cool temperature of the scrotal sac is essential for the proper development of spermatozoa. If the testes of adult rats are returned to the abdominal cavity and fixed there surgically, they atrophy; heat applied locally to the scrotal sac causes damage, which is reversible if the heating is not too severe or prolonged; and grafts of testes transplanted elsewhere in the animal's body survive only if they are placed in a part of the body at a temperature lower than core temperature. An illustration of the damage done by heat to spermatozoan development is given in Fig. 5.4, which shows the marked drop in concentration of spermatozoa in the rete testis fluid of rats and rams, consequent on only a short period of fairly mild heating. From the time required for recovery of the normal concentration (about 60 days) and the known times of the phases of spermatogenesis, it can be deduced that it was the spermatogonia undergoing mitosis during the heating that were killed, and those not dividing were more resistant to the temperature increase.

The scrotal sac is also an important sensor of ambient temperature and can contribute markedly to the control of body temperature as a whole. For example, a pig placed in the slightly cool ambient temperature of 15 °C will sustain its core temperature by shivering and increased oxygen consumption. If only the scrotum is

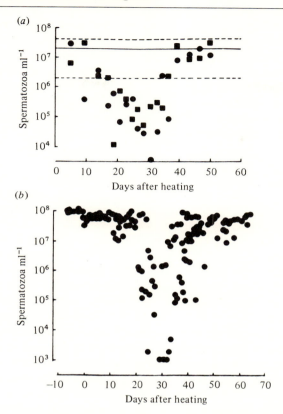

Fig. 5.4. Concentration of spermatozoa in rete testis fluid of (*a*) rats and (*b*) rams at various times after locally heating the testis, the rams to 40 °C for 3 h, the rats to 41 °C for 1 h (●) or 1½ h (■). From Setchell (1978).

then heated, the shivering ceases, and the body temperature consequently declines. Again, if the scrotal sac of a ram is heated to 36 °C, the ram starts to pant, and with continued local warming of this area the panting is so great that its effect lowers deep body temperature. These are illustrations of the way in which the local temperature, that is not of the whole body surface but simply of one small area, can have a great effect on the thermal balance of the whole animal.

Brain

In many mammals, including sheep, goats, cats and dogs, the brain temperature is below the core temperature as measured by the tem-

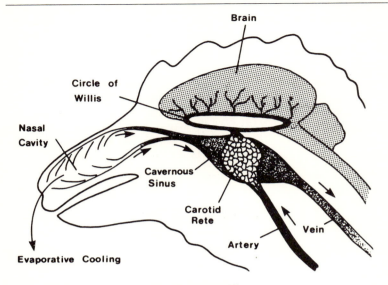

Fig. 5.5. The carotid rete. From Hardy (1979).

perature of carotid blood close to the heart. This difference of temperature is especially marked while the animal is losing body heat by panting, either during exercise or during exposure to a hot environment. For example, it was found that after a high-speed run of 7 minutes, the brain temperature of an antelope was as much as 2.7 °C below that of the carotid blood, which had been heated by the run. This stabilization of the brain temperature in the face of greater fluctuation of core temperature is carried out by a heat exchange device, the carotid rete or rete mirabile, which lies along the course of the carotid arteries just below the base of the brain. Here, the carotid arteries break up into many small branches which are surrounded by venous blood in a pool, the cavernous sinus, or by a network of small veins, the pterygoid plexus. This venous blood is returning from the skin of the face or the upper respiratory tract and is cooled, especially in panting animals, by the airstream passing over the moist respiratory mucous membrane. The arterial blood, on its way up towards the brain via the circle of Willis, is thus slightly pre-cooled (Fig. 5.5).

A similar heat exchange arrangement is present in some birds; the brain is supplied by blood from the two internal carotid arteries which pass through a network, outside the brain, called the rete mirabile ophthalmicum. Here, they come into close contact with a meshwork of small veins returning blood from the back of the head, the eye and upper respiratory passage. Many birds at rest maintain a brain temperature 1 °C below core (or cloacal) temperature; and their brain temperature, like that of mammals, remains more stable than core temperature during exercise or heat exposure.

Even in those mammals (including man and rabbit) without a carotid rete, some pre-cooling of the brain's blood supply during heat stress does occur, by countercurrent exchange with venous blood. The biological significance of this stabilization of brain temperature has not yet been clarified: presumably brain cells are more readily damaged by an elevation of temperature than are the cells of other tissues.

REGULATION OF BODY TEMPERATURE

Regulatory processes

The remarkably small range within which the core temperature is allowed to fluctuate in some mammals when healthy, together with the fact that many infectious disorders have the effect of raising body temperature, has allowed the use of measurement of body temperature as an indicator of health in man and his domestic animals. Such measurements have also provided one of the incentives to research into the mechanisms by which this fine degree of regulation of body temperature is attained.

This regulation can be considered as consisting of three components: a set of sensors which sense ambient or core temperature; a part of the central nervous system which receives and coordinates information from the sensors; and a set of efferent nerves (or hormones) and effector organs by which the body makes appropriate responses. A convenient concept which may help in interpreting some of the observations of thermoregulation is that of a 'set point' temperature, a temperature towards which the core is always tending. If the sensors supply information that the core temperature is drifting upwards above the set point, sweating or panting would be initiated and core temperature would fall; conversely a downward drift from the set point would have the effect of initiating shivering or increasing metabolism. It should be said at once that although the sensory and effector components of thermoregulation are fairly well-known, the task of the central nervous system in overall coordination and control has not yet been worked out, and the meaning of 'set point' in terms of a structure or physiological process is far from clear.

Effectors

The effector side of temperature control has been described, either explicitly or by implication, in Chapters 2 and 4. The effectors would be all those mechanisms by which heat is gained, retained or lost in response to changing environmental conditions, or changing internal states such as generation of heat by feeding and exercise. In summary, these would include: blood vessels of the skin, and their sympathetic innervation; sweat glands and pilo-erector muscles and their sympathetic nerves; respiratory muscles and their motor nerves that bring about panting; the many striated muscles and their nerves responsible for shivering; brown fat, and the nervous and hormonal influences on its metabolism; and the various body tissues which, under influences of hormones such as adrenaline and the thyroid hormones, raise the body's metabolic rate in the cold. These mechanisms can result in great changes in heat production and exchange. Those hormonal changes (induced perhaps by altered day-length) which cause preparations for migration or hibernation might be included among effector mechanisms of temperature regulation. Such a view is questionable, however: the migrating or hibernating animal is not adjusting itself to a change in its thermal environment but simply escaping from it, and migrating or hibernating behaviour probably has as much to do with available food supply as with ambient temperature.

Temperature sensors

The bodies of homeothermic animals have two groups of temperature sensors: the peripheral receptors just below the skin surface, and the deep receptors, most of which are located within the central nervous system, i.e. in the hypothalamus, the medulla and spinal cord. Because of their accessibility the peripheral receptors have been studied more extensively than the central ones. Besides contributing to the mechanism of temperature regulation they also have the function of making the man or animal aware of ambient temperatures: they convey information to the sensory cortex and to 'consciousness'.

The skin receptors are distributed widely over the skin surface and in some animals are particularly abundant on the tongue and

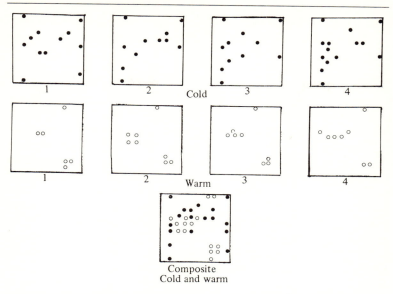

Fig. 5.6. Maps of cold and warm spots on an area 1 cm² on the upper arm. The successive maps were made at intervals of two days. All the spots are combined in the diagram marked 'composite cold and warm'. Redrawn from Dallenbach (1927).

the scrotal sac. They can be mapped by placing the ends of warmed or cooled rods on small areas of skin one at a time. From resultant sensations in man and activity in single afferent nerve fibres in animals, it is possible to locate 'cold spots' which are sensitive to temperatures between 10 and 25 °C, and 'warm spots' which are sensitive to temperatures between 35 and 45 °C. Some parts of the skin are totally insensitive to temperature. Contrary to earlier reports, the 'cold' and 'warm' spots cannot be related to structurally specific organs or afferent nerve terminals immediately below the surface of the spot. Furthermore, the actual position of the spots on the surface is not fixed (Fig. 5.6), so a 'cold end-organ' or a 'warm end-organ' cannot be defined precisely. Single afferent fibres, however, do respond to a limited temperature range. When the tongue of the cat, for example, is covered with water whose temperature is varied, it is possible to record changes in impulse frequency in teased out single afferent nerve fibres in the lingual nerve (Fig. 5.7).

Mapping of spots and the recording of afferent impulses have

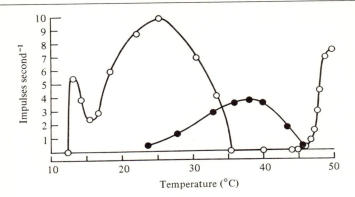

Fig. 5.7. Graph showing the frequency of the steady discharge of a single cold fibre (open circles) and a single warm fibre (closed circles) when the receptors on the surface of a cat's tongue were exposed to temperatures within the range 10 °C to 50 °C. Reproduced from Zotterman (1953), by permission of Annual Reviews Inc., Palo Alto, Calif. USA.

elicited several general principles: (a) cold spots are more numerous than warm spots; (b) the maximum frequency of impulse conduction if the temperature is held steady is about 10 per second; (c) each spot has a characteristic temperature and maximum frequency for this 'stable' response; (d) afferent nerves from cold spots respond to a sudden fall of temperature at the skin surface by a transient burst of impulses of high frequency (about 100 per second), followed by an increased, but much lower steady level of impulses; similarly, afferent nerves from warm spots show transient high frequency discharge when surface temperature is raised, followed by a rise in the steady level of impulses; and (e) these 'transient' responses are elicited mainly around the temperature at which the particular afferent nerve gives its maximum 'stable' response.

Because of these modes of response, and because the various spots have maxima at different temperatures, the temperature sensors of the skin could be providing information both about absolute temperature and about changes of temperature at the skin surface to any possible central regulator mechanism and to 'consciousness'.

Central receptors are more difficult to study. A widely used method is to heat or cool a small carefully localized portion of the central nervous system, by means of a locally placed thermode: this is generally a narrow metallic U-tube through which water of known temperature, slightly above or below that of the tissue, is allowed to run. If the animal makes a thermoregulatory response (panting, shivering, skin vasodilatation) it can be concluded that the thermode is in a brain structure which contains temperature receptors. In this way, for instance, it has been shown that all species examined have temperature receptors in hypothalamic structures on either side of the third cerebral ventricle located in the mid-brain, and in the cervical region of the spinal cord.

Central control mechanisms

In all species of mammals and birds so far studied it seems that the hypothalamus is the principal region in the central nervous system where the afferent pathways from temperature sensors act upon the efferent pathways to thermoregulatory effectors – for example to the vasomotor centre of the medulla – by which autonomic and somatic nerves and endocrine glands make appropriate responses. Experiments which involve placing small lesions in either the anterior or posterior hypothalamus suggest that the anterior region of the hypothalamus is principally involved in the control of responses to a warm environment (panting or sweating, increased skin blood flow), and that the posterior region is principally involved in the control of responses to cold (shivering and other increases in heat production). This apparent dichotomy of controlling influences may relate to the location and densities of the central temperature sensors, warm sensors being more numerous in the anterior hypothalamus. Much of the information about core temperature might thus derive from the direct sensing in the hypothalamus of the temperature of the blood of the carotid artery perfusing this region of the brain, augmented or modified by afferent inputs from peripheral and extra-hypothalamic core temperature sensors.

Efforts have been made to find which of the two sets of sensors are dominant, the central or the peripheral receptors, by giving contrary stimuli to the two sets. It is possible, for instance, to assess a subject's thermoregulatory responses in experimental conditions which cool the skin while warming the blood going to the hypothalamus, or *vice versa*. One such experiment, done by Benzinger (1959) on man is illustrated in Fig. 5.8. The rate of sweating (the response) was measured continuously while the subject sat in a warm room (ambient temperature 45 °C) which would stimulate warm receptors of the skin. The core temperature was then lowered by the swallowing of ice so the central receptors would signal coldness; sweating

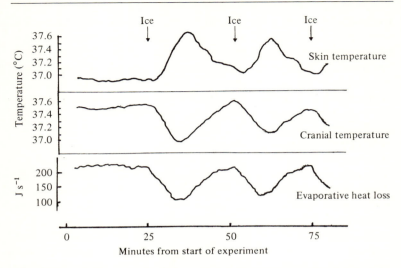

Fig. 5.8. Relation between sweating and brain stem temperature in a warm human subject. Internal temperature measured at the eardrum provides an acceptable index of hypothalamic temperature. From 0–25 minutes, subject equilibrates in warm chamber, 45 °C. Ice ingested at 25, 52 and 75 minutes cools the blood and hence lowers eardrum (hypothalamic) temperature. Notice the parallel decline in evaporation. Skin temperature rises as sweating is inhibited. Redrawn from Benzinger (1959).

was suppressed for about 10 minutes, in spite of the fact that the subject's skin temperature, already high, rose even more during this time because of a reduction in the rate of evaporative cooling. This particular experiment would seem to indicate the dominance of the central sensors over the peripheral sensors in man, but other studies indicate that the relative influence of core and peripheral temperature sensors may vary from species to species, as well as with changes in external thermal insulation which influence the effect of ambient temperature upon peripheral temperature sensors.

Even in man, however, the peripheral sensors are by no means without significance. Their function seems to be to modify the core temperature at which thermoregulatory responses such as sweating or shivering are initiated. For example, for a man in a warm environmental temperature (33 °C) as perceived by the skin receptors, sweating starts when the core temperature is even as low as 36.8 °C, and increases further as core temperature increases; in a cooler environment (29 °C) sweating will not start until core temperature has risen to a much higher threshold, 37.5 °C. There is a similar shift in the core temperature at which shivering occurs with changes in skin temperature. If the skin temperature is cool (say 20 °C) shivering in man starts when core temperature has fallen to 37.1 °C; but if the skin temperature is warm (about 30 °C) shivering will not start until the cranial temperature has fallen as low as 36.5 °C.

These observations have been interpreted as indicative of a 'variable set point' or reference temperature, with which actual hypothalamic temperature is compared; the amount and direction of the mismatching would determine the nature and intensity of the thermoregulatory response. Suppose, for example, that in a thermoneutral environment the 'set point' is 37 °C, and suppose that the actual hypothalamic temperature (determined by the temperature of the cranial blood supply and thus by the balance of heat production and loss of the whole body) is also 37 °C. There would be no mismatch, and therefore no sweating or shivering. If the hypothalamic temperature drops slightly, shivering would start and continue until the additional heat generated has brought the temperature back to the 'set point'; if the blood warms slightly, sweating would start, the body would lose heat and the hypothalamic temperature would be brought down again to the 'set point' of 37 °C. If now, the subject goes into a cool environment, his skin temperature receptors are stimulated, and the effect of this is to raise the 'set point', say to 38 °C. The hypothalamic temperature of 37 °C will now be seen as 'too cold', so shivering would start, and would continue until the hypothalamic temperature is brought up to the new set point, 38 °C. Conversely, in a hot environment, the warm receptors of the skin would lower the 'set point', say to 36 °C. The hypothalamic temperature of 37 °C will now be seen as 'too hot', sweating would start, and the body is cooled, making the hypothalamic temperature fall to this new lower set point.

This idea of a 'variable set point' has been useful in considering other aspects of thermoregulation. It has been supposed, for instance, that the body's defences against microbial infections involve a raising of the 'set point', and that during sleep it is lowered. But it should be remembered that this idea is no more than a concept of how body temperature might be regulated. One ingenious speculation about the meaning of set point temperature is that it is a matter of the balance of the firing-rate of a pair of temperature-sensitive neurones in the hypothalamus, one (A) having a maximal firing-rate at a lower temperature than the other (B), but both cells firing (though not maximally) at some temperature between those of the two maxima. This intermediate temperature is the set point. If the core temperature starts to drift downward, A will fire more, B less, and this *pattern* of firing is the signal for shivering to start. If the core temperature starts to rise, B's firing-rate increases, A's is lessened, and this constitutes the signal for sweating. Afferent impulses from skin receptors or elsewhere impinge on A and B, altering their basic firing-rate and thus the balance between them; this gives the 'variable set point'. This idea is not entirely fanciful; it is indeed possible to record from cells in the hypothalamus whose impulse frequency can be altered by experimental manipulation of temperature. But this idea also has some theoretical objections, and until more experimental work has been done it can be viewed simply as one of a number of pictures designed to help scientists make testable hypotheses about central mechanisms of temperature control.

THE DEAD BAND (NULL-ZONE) OF CORE TEMPERATURE

The various thermoregulatory responses of homeotherms (panting, shivering, sweating, alteration of skin blood-flow and so on) that act as 'effectors' are controlled by the integrated information derived from core and peripheral sensors in such a way as to keep the core temperature stable within fairly narrow limits.

It was shown in Chapter 2 (Fig. 2.6) that when thermoregulatory effector functions of heat production (by shivering or non-shivering thermogenesis) and evaporative heat loss (by sweating or panting) are plotted against *ambient* temperature, one can define a range of ambient temperature – the thermoneutral range – within which there is neither thermoregulatory heat production nor evaporative heat loss, and in which temperature control depends only on manipulation of skin blood-flow by vasomotor nerves. If, instead, heat production and evaporative heat loss are plotted against *core* temperature, one observes for some species a single precise null-*point* of core temperature below which there is augmentation of heat production and above which there is augmentation of evaporative heat loss. Or, in other species, there may be a null-*zone*, within which neither production nor loss is augmented, and in which temperature regulation is carried out only by alteration of skin blood-flow. Again, in other species, there may be a range of overlap within which both sweating and increased heat production are occurring simultaneously at a low level. The camel shows a null-zone: a fairly wide range of core temperature (perhaps 4 °C) within which no thermoregulatory function other than alteration of skin blood-flow is operating (Schmidt-Nielsen, Schmidt-Nielson, Jarnum and Houpt, 1957). In man, in certain ambient conditions, and also in the goat, there may be a null-point (about 37.3 °C for man) or even a small overlap of thermoregulatory heat production and evaporative heat loss (Cabanac and Massonnet, 1977; Jessen, 1977). Even within one species however, the plot of thermoregulatory functions against core temperature will vary according to the prevailing ambient temperature, since this will alter the signals received centrally from peripheral temperature sensors, and these variations will in turn modify the relations between activity in core temperature sensors and thermoregulatory affectors.

CONTROLLED HYPOTHERMIA

In the state of torpor (p. 56) an animal's metabolic rate and body temperature fall. In this condition the energy demand of the tissues is very low, but the state is perfectly compatible with life and the animal emerges from the torpid state each day. An analogous state of low energy demand of the tissues has been used in surgical operations on the heart, which can be carried out only if the heart is empty and motionless. If the energy requirement of the tissues is made very low, the whole cardiovascular system can be interrupted temporarily while the surgical operation is in progress, and then reconnected, allowing the tissues to recover their normal level of metabolism. But whereas in the hypothermia of torpor it is probably the low metabolic rate that produces the fall in body temperature, in the controlled hypothermia of the operating theatre it is the lowering of the body temperature which produces the fall in metabolic rate.

Controlled hypothermia is achieved by placing an anaesthetized patient in a cool room and giving him vasodilator drugs to relax his skin blood vessels. This lowers his temperature to about 28 °C, and the heart is stopped, usually by locally injected potassium. Precooled blood is then perfused through his circulation by an external pump and cools his tissues to 20 °C or even 15 °C. A hypothermia of 15 °C protects the brain cells during complete circulatory arrest, and if the proposed operation takes no longer than an hour, the circulation may be stopped altogether. If the operation takes longer, pre-cooled blood is circulated to the brain and other tissues. Afterwards, the body is rapidly re-warmed to 30 °C by the perfusion of warm blood, and then up to the normal body temperature by external heaters. The invention of this heroic method has permitted numerous operations on hearts either congenitally deformed or damaged by disease, and so has saved or prolonged many human lives.

SUMMARY

The core temperature of various species of homeotherms lies in the range 35 °C to 43 °C, and many species of birds have a range slightly higher than mammals. All species show a daily fluctuation which may be particularly large in certain mammals of hot dry climates. Those species which undergo torpor or hibernation may show a marked fall in body temperature in these states, even becoming temporarily poikilothermic.

The regulation of core temperature requires the integrity of hypo-thalamic mid-brain structures, which contain temperature-respon-sive neurones and which receive the afferent neural input from cold and warm sensors in the skin and other parts of the CNS (medulla and spinal cord). The control mechanism behaves as if a certain 'set point temperature' were being defended, the body's mechanisms of heat production, loss or retention being adjusted accordingly. The meaning of this hypothetical 'set point' is not yet clear. One idea is that the set point is a property of the balance of firing-rates of the hypothalamic temperature-sensors themselves, modifiable by affer-ent impulses.

The body temperature and metabolic rate of humans can be low-ered artificially for brief periods during surgical operations, and restored again without damage to tissues.

CHAPTER SIX

Temperature regulation in the new-born and in old age

THE NEW-BORN

The new-born mammal experiences at birth perhaps the most marked thermal change that it is going to encounter during its lifetime. In the uterus it has developed in an environment characterized by the remarkable degree of temperature stability that is associated with the mature mammal, so although it may be born in a warm nest or a warm room it is immediately exposed to evaporative heat loss, and although it may find itself in a still-air environment it probably also loses heat by natural convection and by radiation. In addition, the new-born's new thermal environment fluctuates, and thermoregulatory responses are called into play.

Except for herd mammals, which are well developed at birth, new-born mammals are generally unable to respond adequately to environmental stresses. This is particularly so in rodents, which are born at a quite immature stage. The more immature a new-born animal appears to be, however, the more tolerant it may be of environmental conditions which result in hypothermia or hypoxia. This tolerance allows it to survive such conditions more easily than can older animals; but being unable to extricate itself from hypothermia, the new-born must instead be rescued by warmth and food provided by the mother. This degree of thermal instability has the advantage of reducing the high metabolic cost of sustaining homeothermy when the ratio of surface area to mass is high.

Body temperature at birth

In the newly born mammal, the deep body temperature falls sharply from its intra-uterine level which is close to that of the mother, and then recovers. How far the temperature falls and how long it takes to recover varies from species to species; findings on the new-born pig are illustrated in Fig. 6.1. Whereas in the pig recovery takes up to a day, or even longer under adverse conditions, in the foal and the calf the drop in temperature is only transient, and in the lamb

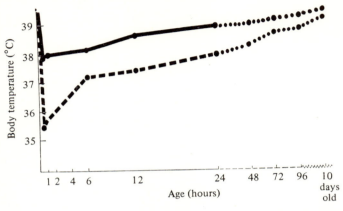

Fig. 6.1. The changes in rectal temperature of the new-born pig in warm and cold environments. Solid line: average of 16 pigs in environment 15 to 21 °C; average birth weight 1.1 kg; broken line: average of 19 pigs in environment −4 to 2 °C; average birth weight 1.1 kg. By permission of the Journal of Animal Science. From Newland, McMillen and Reineke (1952).

recovery occurs in a few hours. In the human infant the body temperature may fall to below 36 °C three hours after birth, but by eight hours it is recovering towards the adult range (Fig. 6.2). Thermoregulation in the immature new-born of the rat is not established until the animals are about 18 days old. The new-born of the cat, dog and rabbit are intermediate between rat and lamb in their development of temperature regulation.

Although exposure to cold increases energy loss, it is also the stimulus that leads to the development of thermoregulatory responses; for example, new-born rats exposed to cold once a day develop thermoregulation more effectively than similar animals kept continually in warm surroundings. Some variation in the environment of the new-born animal, provided that it is not too extreme, can be expected to produce a more hardy animal.

Metabolic rate

Following birth, there is a rise in metabolic rate in the new-born. This is a general phenomenon observed in all mammals that have been examined: it occurs, for example, in man, pig, sheep, monkey and puppy. This rise takes place although the new-born is kept in a

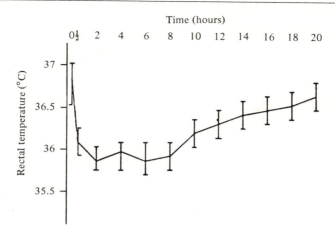

Fig. 6.2. Mean rectal temperatures of 53 new-born infants for 20 hours after normal delivery. The vertical lines show 95% confidence limits. Redrawn from Burnand and Cross (1958).

constant environment, and it is distinct from the rise in metabolic rate that occurs on exposure to cold.

Although all new-born mammals show some rise in metabolic rate when the environmental temperature falls, in the small immature new-born of some species a relatively small degree of environmental cooling can lead to hypothermia. For this reason it was considered for a long time that the new-born rat and mouse are poikilothermic, but this false surmise was due to the fact that below environmental temperatures of about 30 °C the animals were already below their cold tolerance limits (see Fig. 2.6) so the Q_{10} effect was being observed. When measurements were made above 30 °C, the usual homeothermic response of rise in metabolic rate with fall in environmental temperature was observed.

Fig. 6.3 illustrates the metabolic and body temperature responses to thermal environment on the part of the new-born of several species. Amongst the smaller mammals, the cold limit of 10 °C for the guinea pig makes it clearly recognizable as a homeotherm. The lamb has a very low cold limit owing to the combination of a high maximum cold-elicited metabolism (290 W m^{-2}) and a high level of thermal insulation (about 0.4 °C m^{2}W^{-1}). The new-born baby, however, has a high cold limit due to a low maximum metabolic rate (70 W m^{-2}) coupled with a low insulation (0.17 °C m^2 W^{-1}). The

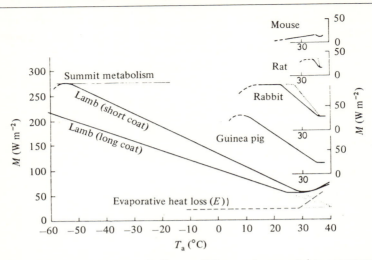

Fig. 6.3. The relation of metabolic rate (M) to environmental temperature (T_a) for the new-born mouse, rat, rabbit and guinea pig, and in new-born lambs with either short or long coats. From Alexander (1975).

maximum sustainable difference between the deep body temperature and the environmental temperature can be calculated for each of these species. Allowing 10% of the total heat loss as evaporative loss, the product of non-evaporative heat loss and thermal insulation (which gives the sustainable temperature difference) is $261 \times 0.4 = 104.4$ °C for the lamb, and $63 \times 0.17 = 10.7$ °C for the human baby. Subtracting this sustainable temperature difference from the deep body temperature (39 °C for sheep, 37 °C for man) gives cold limits of -65 °C for the lamb and 26 °C for the human baby.

There is thus considerable variation in resistance to cold between the new-born of different species, resistance meaning the ability to maintain body temperature. Tolerance of cold, already mentioned, is different because it is essentially the acceptance of hypothermia from which recovery can take place with the mother's help. Although tolerance of cold is most evident amongst those new-born animals that are both small and uninsulated, it also occurs amongst the larger new-born, such as man and pig. The mechanism of huddling, an effective behavioural control of heat loss shown by nestling birds and by the young of many mammalian species which bear litters, has been discussed in Chapter 4.

Metabolic requirements, whether at the minimal rate in the thermoneutral zone or at the higher rate stimulated by cooler temperatures, must of course be met by energy intake, which, for a young mammal, means its mother's milk. Energy of the milk must meet the needs for maintenance, growth, temperature control and movement. From measurement of the energy content of milk of a number of mammalian species, it is noticeable that the highest concentration of fat and energy occurs in the milk of marine mammals and reindeer – species in which the young are born mature, into a very cold environment, and are able to move actively and follow the mother from the moment of birth.

Temperature control in human infants

The naked new-born baby can thermoregulate over the environmental temperature range 36 °C down to 26 °C, at which temperature it reaches its cold limit and has doubled the thermoneutral metabolic rate. The relation between heat production and environmental temperature is shown in Fig. 6.4, from which it can be seen that there is a minimum heat production of about 33 W m^{-2} over the temperature range 32.5 to 36 °C. However, the range corresponding to the difference between peripheral vasoconstriction and vasodilatation is only 32.5 to 33.5 °C, which may represent a more realistic zone of thermal neutrality since above 33.5 °C there is an increase in evaporative heat loss.

Although the human baby has so little thermal insulation, it does have control over peripheral vasomotor tone even on the first day after birth. By this means it can bring about a three-fold change in tissue insulation, although this is only a minor part of the total core-to-environment insulation. The importance of added thermal insulation is emphasized when one realizes that the thermoregulatory range is altered from 36–26 °C to 30–10 °C by simply wrapping the baby in a blanket. Since a large proportion of a baby's heat loss occurs from the head, wrapping round the head is especially important. Examples in other animals also indicate the importance of added insulation. Given adequate nesting material, mice can readily raise their young in nests in cold-stores at an ambient temperature of −4 °C. The young in litter-bearing species huddle together and so diminish heat loss, a facility shared by piglets, puppies and kittens; this is in contrast to the single isolated lamb where isolation or exposure to a demanding environment can lead to a high rate of heat loss and death.

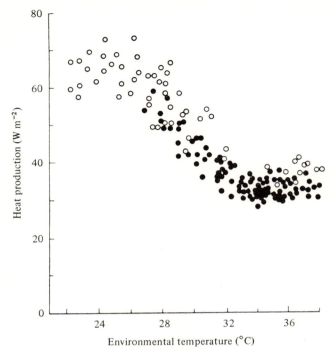

Fig. 6.4. The relation between heat production and environmental temperature in six babies who weighed approximately 2.5 kg when 7–11 days old. Open circles show data obtained during periods of physical activity. From Hey (1974).

The baby's metabolic rate is considerably affected by the level of physical activity and by whether the baby is sleeping. It shows increased activity when subjected to thermal stress, whether this is due to high or low temperatures, and heat production is increased. Metabolic rate per kg remains almost constant during the first year after birth, which suggests a proportionality to body weight, as discussed in Chapter 2.

No one has ever described visible shivering in babies, although the presence of brown fat means that non-shivering thermogenesis can take place (Chapter 2). The baby has deposits of brown fat in the neck, round the kidneys, in the axillae, and in the interscapular region; brown fat may amount to 1% of body weight at birth. The sleeping baby exposed to a cold environment can increase heat

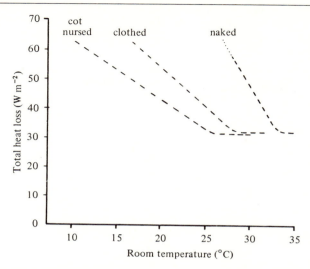

Fig. 6.5. The relation between heat loss and room temperature in a typical baby weighing between 2 and 3 kg when more than 2 days old. Clothing and bedding increase the resistance to heat loss and decrease the temperature necessary to provide thermoneutral conditions. From Hey and O'Connell (1970).

production without shivering or physical activity, and this may be due to the thermogenic activity of brown fat.

There is thought to be a 24-hourly rhythm of body temperature for new-born infants as for adults, but the amplitude of fluctuation is much smaller; mean day-time temperature in infants may be only 0.1 °C higher than night temperature, whereas for adults it may be 0.8 °C higher.

Heat loss in infants

The small size of the new-born animal makes it susceptible to cooling, a tendency that progressively diminishes as it grows, until in the mature animal the susceptibility in many species is to high rather than to low temperatures.

The relation between heat loss and room temperature for a baby weighing 2 to 3 kg is shown in Fig. 6.5. The lines correspond to the heat production line BC of Fig. 2.6, and increasing thermal insulation by clothing and by the protection of a cot makes the slope more shallow, as expected, so reducing the metabolic demand of any

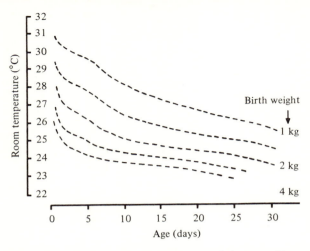

Fig. 6.6. Estimates of the room temperatures needed to provide a cot-nursed baby with conditions of optimum warmth (thermal neutrality) during the first month of life. The baby is assumed to be clothed and wrapped in blankets under draught-free conditions. From Hey and O'Connell (1970).

given room temperature, and simultaneously lowering the critical temperature. For a naked baby weighing 3 kg at birth an undemanding temperature in draught-free surroundings is about 34 °C on the day of birth, falling to 32–33 °C within two days, and then changing only slowly. Temperatures about 1 °C higher than this are necessary for very small babies. Temperatures that are estimated to provide thermally neutral conditions for cot-nursed babies of different weights during the first month after birth are given in Fig. 6.6. In rooms that are colder than 10 °C, babies can become hypothermic; such babies may appear to have a good facial colour, not pale or cyanotic, and they may not cry, but their body temperatures may be lowered and they may need re-warming.

Active sweating occurs in babies born at full term or even three weeks prematurely when the environmental temperature exceeds 34–35 °C and when rectal temperature rises to about 37.2 °C (Fig. 6.7). The rectal temperature at which sweating begins falls during the ten days after birth. About 400 active sweat glands per cm^2 have been found on the thigh, six times as many as in adults, although each gland produces sweat at only a third of the

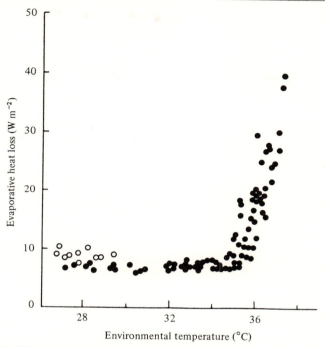

Fig. 6.7. Relation between environmental temperature and evaporative heat loss in babies, when ambient water vapour pressure is about 20 mbar. The subjects were 12 babies, 0–14 days old, weighing between 1.8 and 2.2 kg. Open circles show data obtained during periods of physical activity. From Hey (1974).

maximum rate found in the adult. Sweating has not been observed in babies of less than 210 days post-conceptual age.

Babies in incubators

Babies are normally clothed and placed in a cot, but sometimes, particularly with premature infants, the baby is kept naked in an incubator that is maintained at the appropriate temperature. Apart from important considerations of excessive air movement producing drying and convective heat loss, the use of the incubator provides an example of the importance of taking radiant heat exchange into account in establishing a suitable thermal environment. An important feature of long-wave radiation is that it does not pass through the majority of substances that are transparent to the shorter wave-lengths of the visible spectrum, such as glass and most plastics. This so-called 'glass-house' effect applies to the Perspex (polymethyl methacrylate) walls of a hospital incubator for babies: being opaque to long-wave radiation, the baby will not lose heat by long-wave radiant exchange provided temperatures of the surfaces of the Perspex wall and the mattress on which it lies equal that of the baby's skin. The temperature of the incubator wall is therefore important in determining the baby's radiant heat loss, because the wall occupies the major part of the solid angle subtended at the baby. The incubator wall temperature actually lies nearly midway between the temperatures of the incubator air and of the room air, so that in a cool room the baby's radiant heat loss can be considerable. A thin Perspex shield placed between the baby and the incubator wall will reduce this loss. The shield tends towards the temperature of the incubator air and, since it is opaque to long-wave radiation from the baby, it acts as a warmer radiant environment than the incubator wall. The use of such a radiant heat shield significantly lowers the thermal demand of a baby's environment as shown by a reduction of oxygen consumption rate from 9.0 to 7.6 ml kg^{-1} min^{-1} (mean result from six babies). The shield also has the effect of raising the baby's skin temperature about 0.5 °C as shown in Fig. 6.8.

OLD AGE

During the last 20 years, the condition of hypothermia (lowered deep body temperature) in elderly people has been recognized, and indeed some sufferers have required hospital treatment to return the body temperature to the normal range. The lower limit of normality is usually considered to be 35.5 °C. A figure below this, for deep body temperature, would indicate incipient hypothermia. In a survey of 1000 people aged 70 and above, about 10% were found to have a deep body temperature (as measured by urine temperature) of 35.5 °C or below, and there was a negative correlation of temperature with age: that is, the older, the colder. There appear to be at least two reasons for this. One is the lowered metabolic rate of the elderly: there is simply less heat production in the metabolizing tissues. The other is the lowered ability of old people to perceive changes in ambient temperature and to make the appropriate behavioural response. This was clearly demonstrated by Collins, Exton-Smith and Doré (1981) in an experiment in which two people, one

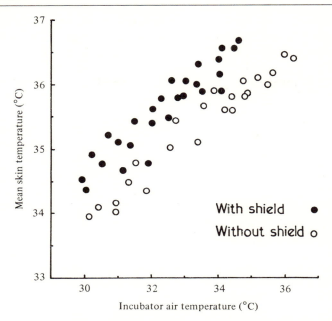

Fig. 6.8. The effect of a radiant heat shield on the mean skin temperature of 13 healthy human infants in an incubator over a range of incubator air temperatures. Room temperatures 20–22 °C. From Hey and Mount (1967).

aged 24 and one aged 70, were each in turn asked to sit for about half an hour in a room at 19 °C (cool but not uncomfortably cold) and then to adjust an electric heater to give the room their preferred comfortable temperature. The heater had no thermostat and had therefore to be adjusted up or down at frequent intervals, and the subjects were questioned every 15 minutes about their state of comfort. The mean preferred room temperature for the two subjects was the same, 23 °C; but the older man allowed the ambient temperature to make much greater fluctuations, and made fewer adjustments to the heater, while attaining his preferred ambient temperature (Fig. 6.9). Similar results were obtained in the whole group of 17 elderly men, aged 70 and upwards, who were compared in this experiment with men in the 18–39 years age range. This observation explains the finding of another population survey of people aged 65 and older, that there were no complaints of discomfort even in people who had incipient hypothermia: the coldest people had the

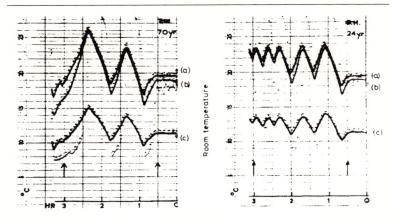

Fig. 6.9. Room temperature controlled by elderly subject (man aged 70), left; and by young adult (man aged 24), right. Room temperature maintained at 19 °C for 30 minutes before remote control period. Time scale from *right* to *left*, 0–3 hours. Air temperature measured at 2 m from floor (a), and at table height (b); (c): wet bulb temperature. From Collins, Exton-Smith and Doré (1981).

least awareness of discomfort from cold. This lack of perception, together with the low metabolic rate, obviously puts old people at risk.

A third possible factor in the hypothermia of the elderly is that the autonomic processes by which both vasomotor tone and sweating are controlled may become progressively less efficient. An impaired response to a cool environment by the usual peripheral vasoconstriction would mean a failure to maintain a steep temperature gradient between core and skin.

Although there is a commonly held view that the hypothermia of old age is related to poverty and poor social conditions, it is difficult to find clear evidence for this. One piece of evidence, which arose from a survey of people of 70 years and upwards (Fox, Woodward, Exton-Smith, Green, Donnison and Wicks, 1973), concerned a small subgroup (about 100 people) of this group in whom the body temperature was found to be below 35.5 °C, the lower limit of normality. In this subgroup, half the subjects were receiving supplementary pensions, that is, they were recognized as being in financial need. Of those whose temperatures were in the normal range, 36 °C and above (a group of about 700 people), only a third were receiving supplementary pensions. This difference was significant ($p < 0.01$).

SUMMARY

The new-born of most mammalian species need protection during their first post-natal hours or weeks, when for the first time they have to maintain a body temperature higher than that of their environment; this is particularly so for the human baby. Although they are able to make regulatory responses (increased metabolic rate, vasomotor adjustments, sweating) the new-born of many animals have little tissue insulation, and their thermoneutral range is at a much higher ambient temperature than that of mature animals. In nearly all species, additional insulation – such as that provided by a nest, litter-mates or blankets – is needed to help them through this vulnerable period.

The elderly human also needs protection from cold, but for a different reason. In old age he becomes less able to perceive the fluctuations of ambient temperature, to generate the necessary metabolic heat in cold environments, and perhaps also to make the required vasomotor responses for controlling deep body temperature.

CHAPTER SEVEN

Adaptation, acclimation and the climate for comfort

CLIMATES OF THE EARTH'S SURFACE

The homeothermic species (mammals and birds) have colonized almost all the climatic regions of the earth to some extent. Their spread has been determined and limited partly by their ability to maintain a certain range of deep body temperature in the face of extremes of heat and cold, and partly also by the available food supply, which in the long run is dependent on plant life.

The climates of the earth can be placed in four groups: (1) arctic and subarctic, where the homeotherm's problems are the cold and the short summer growing season; (2) the temperate, where there is a marked seasonal variation so the homeotherm must be able to live at, say, $-5\,°C$ in winter and $30\,°C$ in summer; (3) dry tropical and subtropical, which may show considerable daily temperature variation, say $0–40\,°C$, but is satisfactory as to ambient temperature provided that the water supply permits adequate evaporative heat loss at the warm time of day; and (4) wet tropical or rain forest, where the environmental temperature shows little daily or seasonal variation and the main problem as to heat balance is the high relative humidity, making evaporative heat loss difficult.

Climatic conditions on land surfaces vary with altitude as well as latitude. At 5000 m on the equator on Kilimanjaro conditions are similar to those of the low altitude arctic tundra; and at sea level in temperate regions, say in S.W. Britain in places sheltered from wind, the conditions are close enough to the subtropical to allow growth of Mediterranean species of plants. Oceans show less extreme variations with latitude and with season than do land surfaces, and for this reason islands and small land masses and the edges of continents – places exposed to ocean winds – have much less extreme seasonal variations of climate than do the centres of continents.

DEFINITIONS OF TERMS

The terms 'adaptation' and 'acclimatization' are used loosely in everyday speech: '... he has adapted his life-style to that of his neighbours ...'; '... he is becoming acclimatized to urban life ...'. The terms are also used in several different senses by biologists. It therefore becomes very important to define the words as they will be used in the present context. In this chapter 'adaptation' means a feature of a species or of an individual animal – structural, physiological or behavioural – which allows it to survive and reproduce under apparently adverse conditions, or when conditions change, or when the animal itself moves from one set of conditions to another. 'Acclimation' or 'acclimatization' is one particular form of adaptation – the form which allows survival in adverse or varying climatic conditions. Such adaptations may be elicited in the individual as a response to the climatic conditions, or selected in the species by the climatic conditions. However, an animal may show an adaptive change which actually precedes the environmental change which gives the adaptation its biological advantage: such an adaptation could not possibly be a response to the environmental condition. Again, not all responses to climatic conditions have survival value. It is only when the structure, physiological device or behaviour pattern can clearly be shown to be biologically advantageous that it can be described as an adaptation.

ADAPTATION TO CLIMATE

Arctic and subarctic climate

Several characteristics of structure and behaviour have enabled certain species and groups of homeotherms to spread to and survive in polar regions. Among these features are (1) the thick tissue insulation provided by the subcutaneous fat or the coat or pelt which slows down heat loss by all non-evaporative routes; (2) large size, which gives a low surface-to-volume ratio and thus limits the surface from which heat can be lost; and (3) use of marine or semi-marine habitat, which gives less annual fluctuation of environmental temperature than that experienced by many fully terrestrial species. Although water has much greater thermal conductivity than air, marine species escape the wind chill of the arctic land masses. Species such as whales, seals, polar bears and penguins show some or all of these three features.

Insulation has been studied in several species of seals and sea lions, and also in husky dogs. In harbour seals (*Phoca vitulina*) the thin hair provides very little insulation in water because it becomes wetted, so the skin temperature is almost at the same temperature as the surrounding water. In ice-cold water, however, the 'shell' of cool tissue surrounding the inner warm 'core' of the body may be 40–60 mm deep, giving a thick layer through which heat from the core must pass, whereas in warm air, the cool shell is only about 20 mm deep. This alteration of shell thickness is carried out by altered vasomotor tone, changing the vascular bed. The fur-seal (*Cattorhinas ursinus*) and the sea otter (*Enhydra lutis*) have water-repellent fur. It provides such good insulation that (for the fur-seal) the temperature of the skin below the fur may be nearly as high as the core temperature. When such an animal moves from ice-cold water onto land where the air temperature may be far lower, water is quickly shed from the surface of the fur before it freezes solid. The thick fur of the Eskimo dog provides such good insulation that the dog can sleep curled up on snow at −20 °C without a fall of core temperature.

The thick layer of insulation of arctic animals does not prevent their being able to survive in warmer environments. Various species of whales and dolphins are found in tropical as well as arctic seas, and indeed a single individual may roam widely between latitudes. The semi-aquatic mink, nutria or coypu (*Myocastor coypu*), well-known for the thickness of its pelt, occurs from the subarctic regions of northern Europe and Asia to the subtropical parts of the North American continent. Seals and sea lions are found on tropical as well as arctic shores. When a sea lion is on land in a warm climate heat loss becomes its thermal problem, which it solves by waving its highly vascular flippers, and by throwing wet sand or shingle over its body. Polar bears, though good swimmers, soon become over-heated if they run on land, their thick insulation preventing much dry-land exertion; a man can easily outstrip a polar bear, which simply stops moving when its core over-heats. Nevertheless, polar bears and many other arctic species appear to live in reasonable comfort out-of-doors in zoos, in temperate and tropical climates.

Many subarctic species grow additional fur, thus thickening the coat during the autumn. It is interesting that this thickening of the coat is a response, not to a cold ambient temperature, but to the shortening day-length of autumn: there must be a hormonal rhythm involved. The effect of the response precedes the physiological requirement for it – a very successful climatic adaptation. Indeed, considering the time (days or weeks) required for this thickening, a growth of fur which occurred only after, and in response to, a lowering of the ambient temperature in winter would be of a much reduced biological value.

Although humans often increase body weight and body fat, including subcutaneous fat, during the winter or during visits to polar regions, no biological value of these changes has been demonstrated. The increased body weight of winter is likely to be associated with the increased appetite arising from a cooler environment together with a limitation of exercise. So, although metabolic rate and heat output may have increased in winter, the additional energy intake together with diminished exercise result in temporary energy storage.

People have noticed that during residence in arctic regions their nails grow more slowly, a consequence of decreased skin blood flow resulting from the raised vasomotor tone associated with the cool conditions. This slow growth of the nails, though resulting from the low ambient temperature, could hardly be considered as an adaptation, since it is highly unlikely that it has any biological advantage. It is mentioned here merely as a warning against over-hasty interpretation of apparent 'adaptations' in teleological terms.

A remarkable adaptation to cold has been observed in the leg of the herring gull (*Larus argentus*). As has been indicated in Chapter 4, the foot and lower part of the gull's leg may fall to a temperature close to 0 °C, the countercurrent heat exchange system of the blood vessels retaining heat in the more central region of the body. The peroneal nerve arises from the lower part of the spinal cord, and thus runs for part of its length through the tibial part of the leg, which is well covered with muscle tissues and feathers and hence has a temperature close to that of the core; it also runs through the metatarsal part of the leg, which being free of feathers and having minimal muscle is at a much lower temperature. The fibres of this nerve show different properties along their length: in the tibial (warm) region, impulse conduction will fail if the ambient temperature falls to about 8 °C, whereas in the metatarsal (cold) region the fibres are more resistant to cooling and continue to conduct at 0 °C. Furthermore, the difference in properties between the two regions of the nerve is greater in gulls which have been kept before slaughter at 0 °C (cold-adapted) than in those kept at 24–35 °C (heat-adapted). This is a rare example of a cold-enhanced climatic adaptation occurring not as an anatomical

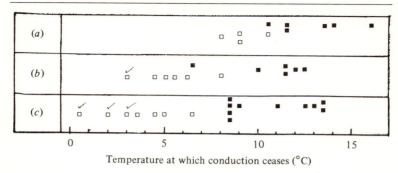

Fig. 7.1. Temperatures during experimental cooling of the peroneal (leg) nerve of the herring gull (*Larus argintatus*) at which the conduction of nerve impulses ceased. ■: tibial (upper) section of the nerve; □: metatarsal (lower) section of the nerve. A tick over a symbol means that the nerve section was still conducting at this temperature but was not cooled further for fear of irreversible damage. (*a*) nerves from 'hot-adapted' gulls which had been exposed to 24 °C to 35 °C; (*b*): nerves from 'warm-adapted' gulls which had been exposed to 26 °C to 30 °C; (*c*): nerves from 'cold-adapted' gulls which had been exposed to − 1 °C to 6 °C. Redrawn from Chatfield, Lyman and Irving (1953).

structure, a physiological device or a behaviour pattern, but within the confines of one single cell (Fig. 7.1).

Temperate climates

The property required of the thermoregulatory systems of temperate zone mammals and birds is versatility. The animal must adapt to the very marked annual cycle of temperature, ranging from about 0–5 °C to about 30 °C. Most species respond in two ways, one physiological, one behavioural. The first is the extra thermoregulatory heat production in winter, as ambient temperature falls below the critical temperature for the species, and less heat production in summer. Many of the mammalian species of temperate zones have critical temperatures between 17 and 27 °C, about the middle of the range of the annual cycle of ambient temperature, and thus metabolic rate is readily adjustable in the course of the year. Arctic and subarctic species have a lower critical temperature: some species do not call on extra thermoregulatory heat production until the ambient temperature has fallen well below 0 °C.

The second climatic adjustment to the temperate zone is a marked annual cycle of activity. Thus all species of birds and many species of mammals raise their young during spring and summer, when food is most abundant and the climatic demand in energy is least. The success of these adjustments is shown by the enormous number of homeothermic species of all sizes which inhabit the temperate zones of the earth.

Dry hot climate

The dry hot parts of the earth include the bush and grassland areas of most of the African continent, of southern and eastern Australia, and of the southern and western part of the North American continent, as well as the real desert regions – Sahara, central Australia, and Mojave among others. A dry hot environment is not particularly demanding for the homeotherm, provided that it can obtain adequate water from food, drink or metabolic water production to balance its inevitable evaporative water loss. Many physical anthropologists consider that man evolved in just such a subtropical climate. A vast number of mammals and birds live and breed successfully in the African bush. The survival of many species depends on the availability of water holes, rivers or lakes. A number of mammals are most active at dusk or at night, spending the day lying in the shade of trees or rocks, or in burrows in the case of smaller species, a behaviour pattern which reduces the radiant heat load. The extremely arid deserts support very little life of any kind and those few species living there have had to solve the problems of limited food and water supplies as well as of the heat load. They usually live a nocturnal life, and derive most of their water metabolically from the oxidation of the sparse vegetable food.

A breed of sheep, the merino, imported into the hot semi-arid regions of Australia, flourishes in that environment and apparently grows a fleece even longer than that of merinos in Spain, the place of origin of the breed. It seems likely that this growth is the consequence of the warm environment in that the skin blood flow and thus skin nutrition would always be large. From the dire thermal consequences of shearing sheep in a Queensland summer, it would seem very likely that the additional length of the fleece has a thermoregulatory function in reducing the animal's radiant heat load. Thus fleece length could also be an 'adaptive' response.

Warm wet climate

The warm wet climate of the tropical rain forest of South America, West Africa, New Guinea and Malaysia is remarkably uniform, showing little annual or 24-hourly change. It is not a demanding environment for the homeotherm, provided that its food seeking does not require intense or sustained activity. The main thermal problem is heat loss into an atmosphere high in both temperature and relative humidity. The slow movements of such animals as sloths and armadillos do not generate much heat. Birds and the many species of monkeys, largely arboreal, are able to make use of any breeze available in the tree-tops in assisting evaporative heat loss.

ACCLIMATION

Numerous experiments have been made in which animals or humans have been subjected to specific ambient temperatures in controlled conditions. The set of responses which follow is known collectively as acclimation. They may be, but are not necessarily, adaptations to the temperature. Two examples of such controlled exposure to extremes of ambient temperature will be given.

In the first example (Dauncey, Ingram, Walters and Legge, 1983), two litter-mate pigs were taken at 2 weeks of age for rearing in two temperature-controlled rooms, one at 10 °C ('cold pig') and one at 35 °C ('warm pig'). The cold pig was fed *ad libitum*; the warm pig was fed on an amount of food limited so as to keep its body weight as close as possible to that of its cold litter-mate. Both pigs grew to a weight of 11.3 kg at 8 weeks. In appearance, however, they were completely different at this age (Fig. 7.2). The warm pig had only a slight hair covering, a pink skin, and the proportions of the typical adult pig. The cold pig had a short thick snout, limbs and tail, a white skin and a heavy hair growth. Subsequent histological study of the injected skin revealed that the skin of the cold pig contained far fewer capillaries than that of the warm pig.

Of these morphological responses to the different environmental temperatures, the only one which could be considered an adaptation is the greater hair growth of the cold pig, which might have provided more thermal insulation (though its thermal insulation was not specifically measured). The cold pig's slow proportionate growth of the extremities and lack of skin capillaries were probably the direct response to the intense peripheral vasoconstriction consequent on the cool ambient temperature, which would cause a poor nutrient supply to the extremities during growth; this is precisely analogous to the slow growth of fingernails of men living in polar regions. The

Fig. 7.2. Litter-mate pigs exposed to different ambient temperatures from 14 days to 56 days age, but fed in such a way that they had the same increase of body weight in this time. Pig on left, kept at ambient temperature 35 °C. Pig on right kept at ambient temperature 10 °C. By courtesy of Dr D.L. Ingram.

heavy hair growth, however, must have occurred in spite of the poor skin blood supply.

The second example of an experiment in acclimation (Edholm, Bedford, Ellis and Mackworth, 1960) concerns a group of 12 healthy young men subjected to daily periods in a warm room kept at a temperature of 37.8 °C dry-bulb, 34.4 °C wet-bulb, with air speed of 0.73 m s^{-1}. While in the room, the men alternated work and rest periods; the exposure continued for 12 days. By the 7th day (Fig. 7.3) the mean standing heart rate of the subjects, which had been 114–120 beats per minute on Day 1, had fallen to 105 beats per minute. By the 9th day, the sweating rate was 32 g kg^{-1} body weight – a rise from the initial level of 24 g kg^{-1}. By the final day the rectal temperature, which had been 39.4 °C (103 °F) on Day 1, was down to 37.5 °C (99.5 °F). From similar observations in numerous experiments of this kind it is apparent that humans can be acclimated to heat, the most conspicuous adjustments being a lowering of the

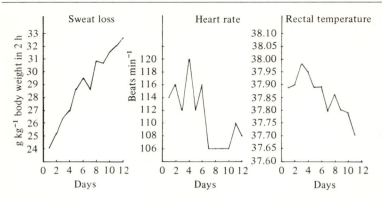

Fig. 7.3. Mean sweat loss, heart rate and rectal temperature of 12 young
men exposed in a warm climatic chamber for varying periods for 12 days.
Days 1–3, 2 h exposure; days 4–6, 3 h exposure; days 7–12, 4 h exposure.
In the chamber, the conditions were 37.8 °C dry-bulb, 34.4 °C wet-bulb,
0.73 m s^{-1} wind speed. The young men, all of whom had lived in the
tropics for at least 15 months before the experiment, spent the time in the
climatic chamber alternating ten minutes' work (step test) with 20 minutes'
rest. Sweat loss measured in g per kg body weight, over the first 2 h; heart
rate measured (standing) after the 5th work period; rectal temperature
measured after first 2 h. Data from Edholm, Bedford, Ellis and
Mackworth (1960).

heart rate, an increase in volume of sweat and a lowering of the
threshold skin temperature at which sweating starts. Another important adjustment is that the sweat, after acclimation, contains a
lower concentration of sodium chloride, the reabsorption of salt in
the ducts of the sweat glands being due to a high circulating level of
the hormone aldosterone.

The success with which man can acclimate to heat has made it
possible to pre-acclimate people about to undertake work in a tropical setting. There is also some evidence of men being able to acclimate to cold. Arterial blood pressure may rise temporarily under the
stress of an arctic environment, then gradually return to its normal
level. Men out on expeditions in arctic conditions may at first shiver
and feel much discomfort when trying to sleep at night. After a time
they show either 'metabolic' or 'hypothermic' acclimation. In
metabolic acclimation the metabolic rate is raised by non-shivering
thermogenesis and so the core and peripheral temperatures remain
high during sleep and shivering is prevented. In hypothermic accli-

mation, the core temperature is allowed to fall – an exaggeration of the lowered set point which normally occurs in sleep; again, shivering is inhibited. Certain races (e.g. aborigines of central Australia, and Indians of Tierra del Fuego) who use little clothing and shelter show this nocturnal lowering of core temperature to a considerable extent. However, much of the increasing degree of comfort that people seem to experience after living for some weeks in a cold place is probably a matter of psychological habituation rather than physiological adjustment.

THE CLIMATE FOR COMFORT

The efficiency of people depends greatly on their thermal comfort, and man's high technology enables him to manipulate his immediate environment to maximize this comfort. It is therefore important to know what are the conditions for comfort, and, when work has to be performed out-of-doors, what limits can be expected for human performance.

The variables on which man's thermal comfort depend are wet- and dry-bulb ambient temperature, relative humidity, and wind speed. Environmental physiologists concerned with work loads of men in different climates, or with building and engineering, have drawn tables or diagrams which allow optimum conditions to be predicted. Examples are given in Figs. 7.4 and 7.5, which have been drawn from data derived from experiments on human subjects or model objects in climatic chambers and wind-tunnels, and verified in field conditions such as those of the Sahara, South African mines, or the arctic. Fig. 7.4 relates dry-bulb temperature to absolute and relative humidity, but omits wind speed, the other important climatic variable. It shows that for people accustomed to a temperate climate, dry-bulb temperatures in the range 12 °C to 34 °C are tolerable; a relative humidity as high as 80% is tolerable if the dry-bulb temperature is no higher than 20 °C, but at higher temperatures the tolerable relative humidity falls off quite sharply. Those people who are acclimated to tropical conditions however can tolerate a 70% relative humidity even at an air temperature of 27 °C. Fig. 7.5 relates wind speed to air temperature in arctic and subarctic conditions. It shows, for instance, that at 10 °C a wind speed of 12 m s^{-1} (nearly 30 m.p.h.) would be tolerable, but if the temperature falls to 0 °C a wind speed of only 3 m s^{-1} (about 7 m.p.h.) would be felt as very cold (II on graph). Diagrams such as Figs. 7.4 and 7.5 allow

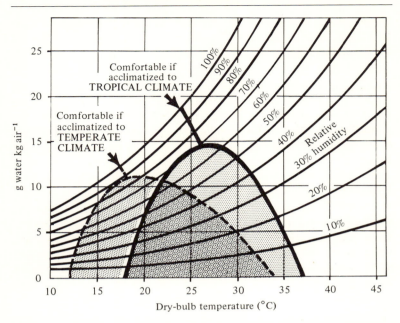

Fig. 7.4. The zone of comfort, in relation to temperature and humidity, for people adapted to temperate and to tropical conditions. From Folk (1974).

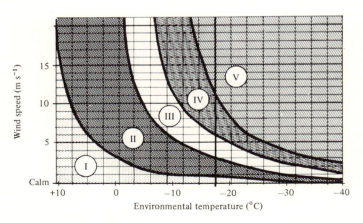

Fig. 7.5. The conditions of wind and temperature, showing zones of increasing severity: I: comfort with normal precautions; II: very cold; III: bitterly cold; IV: freezing of human flesh begins; V: exposed flesh freezes in less than one minute. From Folk (1974).

predictions of human tolerance to adverse climates and are of great assistance to medical advisers in charge of people working in field conditions in extremes of heat and cold.

SUMMARY

All four main climatic regions of the earth provide habitats for mammals and birds. Many arctic and subarctic species have thick external or tissue insulation. Species of temperate regions adjust behaviourally and by greater or lesser heat production to the annual cycle of climatic variation. Dry tropical climates are suitable for homeotherms provided that the water supply is adequate, and the wet tropical environment is also suitable provided that the animal's way of life does not demand much physical exertion. Many species show acclimative changes when subjected to variations in ambient temperature in controlled conditions. Man can acclimate to heat, and to some extent to cold. The definition of the climate for comfort is of practical importance to those responsible for human well-being.

CHAPTER EIGHT

Energy balance and body composition

ENERGY EQUILIBRIUM AND WEIGHT BALANCE

We have seen that temperature equilibrium is achieved if (and only if) heat loss from the body exactly balances the heat generated within it. Chapter 5 indicated some of the processes by which this balance is maintained. Heat cannot be stored in the body in the form of a rise in temperature, except very temporarily during hard muscular exercise, or in certain desert mammals in which a transient rise of several degrees is possible in a hot ambient temperature in day-time, and is followed by a corresponding fall at night.

This chapter is concerned with another aspect of an animal's energy exchange with its environment: its intake of chemically bound energy in the form of food, and output of metabolically freed energy in the form of heat and work on the environment. We shall see that, though storage of freed energy is exceptional since it means a raised body temperature, storage of chemically bound chemical energy as increased tissue mass is a frequent and normal occurrence. Living organisms are of course subject to the Law of Conservation of Energy. An animal is in complete energy equilibrium if, over a period of time, its energy intake and output are identical. This is the state of many adult animals including man. The young growing animal or the pregnant female is in positive energy balance: intake exceeds output, the balance being stored as additional tissue of the animal or the fetus. The starving or near-starving animal is obviously in negative energy balance: output exceeds intake, the balance being made up by the loss of energy due to metabolism of the animal's own tissues, with resulting loss of weight. A satisfactory state in an adult human is the condition, shown by many adults, of constancy of body weight over a span of many years, say, the 20-year span from age 25 to age 45, remaining at about 72 kg (158 lbs) for a man throughout this time. Even this constancy of body weight, however, may not indicate precise energy equilibrium over this time span, because composition of the body may have changed slightly.

If, for instance, 1 kg of fat had replaced 1 kg of muscle, the energy content of the body would have increased slightly without weight change, fat having a high energy content per gram; so the person must have been in slight positive energy balance over the years. There are of course predictable short-term fluctuations of the energy content of most adult animals which in the long-term are preserving an energy equilibrium. Diurnal animals gain energy and weight in the day-time while they are feeding, and lose it at night, and a human's energy content obviously fluctuates with meal-times. The rhythm of energy changes may be annual, in hibernating or migratory species. Such an animal eats enormously and retains energy just before the hibernation or migration, and subsequently loses the stored energy, so its weight and energy content might be much the same on a given date each year.

The adult animal can to some extent preserve its energy equilibrium in the face of disturbances caused by external circumstances. A human consciously limits his energy output when food is in short supply. British prisoners-of-war, on short rations, sleeping in a dormitory on an upper floor, reduced the number of times a day they needed to walk upstairs. But this limitation of energy output is unlikely to be an adequate adjustment and weight loss is unavoidable in the end. Conversely, if food is readily available, an animal which takes more exercise is able to keep its weight constant by eating more. This point can be illustrated by the experiment of which the results are shown in Fig 8.1. Rats ran in a revolving wheel at intervals for different periods of total time each day, from over five hours daily down to only one hour daily. Within these limits, the rats were able to maintain a steady body weight in the face of the varying levels of energy output, by eating either more or less food. The lower limit of weight constancy, one hour of exercise daily, is of some interest. The experimenters wrote . . . 'As work-output is reduced, there comes a point at which food-intake is *not* reduced in line with work.' This may be called the 'sedentary range'. The sedentary rat eats about 230 kJ of food daily regardless of its work-output, and, if its exercise is zero, it puts on weight – a positive energy balance. This is the reason for deliberately limiting the amount of exercise available to commercial livestock such as pigs and geese: the less the movement, the quicker the fattening. These examples of semi-starvation, varied exercise, or exercise limitation, are imposed by external circumstances. The physiological processes of an animal's body would rarely be required to make adjustments

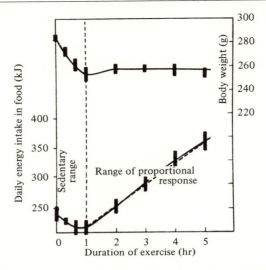

Fig. 8.1. Voluntary food intake (lower line) and body weight (upper line) in relation to exercise in normal rats. Redrawn from Mayer and Bullen (1960).

to such conditions, and energy equilibrium is the normal state of the adult human or animal in the wild.

ENERGY INTAKE AND ITS MEASUREMENT

The amount of energy taken into the body in the form of food may readily be measured. Its measurement is of practical importance to human and animal nutritionists, as well as to food-technologists, experimental food-crop breeders and the like. The measurement is carried out by bomb calorimetry. A small known weight of the well-mixed food is burnt in oxygen under pressure in a heavy well-sealed cylinder (the bomb), the firing being done by completing an electrical circuit in a thin platinum wire passing through the sample. The heat of oxidation is transferred through the wall of the bomb to a known weight of water forming a well-stirred water-jacket surrounding the bomb and insulated on the outside. The rise of temperature in this water indicates the heat generated by complete oxidation of a known weight of dry food, allowance being made for the thermal capacity of the bomb itself. Fats have an energy concentration of about 39 kJ per g dry weight, carbohydrates have about

18 kJ per g dry weight and protein about 24 kJ per g, exact figures depending on the precise nature of the foodstuff concerned. The energy content in terms of kJ per g of average samples of the common human and animal foods is well-known and can be found from published food-tables. But for accurate study of energy balance of man or animals, bomb calorimetry is carried out on samples of the actual foods eaten by the experimental subjects.

The energy content of the food eaten is the 'gross energy intake'. A large proportion of this energy is absorbed in the alimentary canal, but not all: the energy of the residue, passed as faeces from the rectum, must be subtracted from the gross energy intake to give the 'digestible energy intake'. This may be between 75 and 90% of the gross energy intake, depending on species, age, nature of the food and other conditions. Furthermore, within the body many foodstuffs do not undergo complete oxidation analogous with that in the calorimeter. Their residues are passed out from the body, in the form, for instance, of urea, organic acids or nitrogenous bases, which still have some energy content. So the intake, in the sense of energy available to the animal for its metabolism, must make allowance for this. By subtracting faecal and urinary energy output from gross energy intake one obtains 'metabolizable energy intake'. It is this figure which is of most practical value, both in human nutrition and in animal husbandry.

The following example is taken from a day's feed of an adult pig, body weight 35 kg, at an environmental temperature of 20 °C, being fed at about three times its maintenance intake, i.e. undergoing fattening.

Food (A). Amount fed = 1695 g; dry matter concn 88.47%
∴ dry matter intake = (88.47)/100 × 1695 = 1500 g
Energy concn = 17.9 kJ per g dry matter
∴ gross energy intake = 17.9 × 1500 = **26 850 kJ** per day
Faeces (B). Weight of faeces = 998 g; dry matter concn = 33.4%
∴ weight of dry matter = 333 g
Energy concn = 17.1 kJ per g dry matter
∴ faecal energy output = 17.1 × 333 = **5694 kJ** per day
Urine (C). Weight of urine 2545 g; dry matter concn 3.1%
∴ weight of dry matter = 79 g
Energy concn 9.28 kJ per g dry matter
Urinary energy output = 9.28 × 79 = **733 kJ** per day
Gross energy intake (A) 26 850 kJ

Digestible energy intake $(A - B) = D = 26\ 850 - 5694 =$
   $21\ 156$ kJ $= 78.8\%$ of A.
Metabolizable energy intake $= A - (B + C) = 26\ 850 - 6427 =$
   $20\ 423$ kJ per day
which is 76.1% of A; and 96.5% of D.

These figures are typical of mammals including man: the energy
actually metabolized within the body is only about 76% of that
taken in as food, but about 96% of that which is digested and
absorbed in the alimentary canal.

So much for the energy sources of the body as a whole. This is
normally a mixture of fats, carbohydrates and proteins, and this
mixture is used by the body in maintenance of the integrity of the
tissues, deposition of new tissue in the growing or pregnant animal,
resting metabolism (of which maintenance of body temperature is a
major component in the homeotherm) and the performance of external work. All three main foodstuffs are used in synthetic processes of growth and reproduction. It seems that for temperature
maintenance and the performance of work, a mixture of fat and
carbohydrate is used as the immediate energy source. But, because
of the ability of the liver to inter-convert the primary foodstuffs,
attribution of types of foodstuffs to this or that role in the body may
be somewhat unrealistic.

ENERGY OUTPUT

The output of energy from the body, apart from the small amount
associated with the excreta, takes the form of (1) heat loss and (2)
performance of external work. Heat loss, as explained in Chapter 3,
takes place by various routes and its amount depends greatly on
environmental conditions; homeotherms use their energy intake to
make good the heat loss. In general, it is true to say that for most
homeotherms, most of the time, their *resting* metabolism balances
and is determined by their heat loss.

Although most of the energy output is dissipated as heat within
the organism or into the environment, the performance of external
work is a different matter. It is totally irregular and unpredictable,
and is not related to the environmental conditions. Except for those
primitive animals which move on water- or air-currents, all animals
perform external work when moving their own bodies, by walking,
running, swimming, hopping or flying. Indeed even the process of
standing still requires isometric work on the part of the leg and back

muscles. In addition, many species move other objects around with them: mothers carry young, birds carry nest material, squirrels carry nuts. Man, by making and using tools, has enormously increased his ability to move himself and to manipulate other objects. Movement, whether in humans or other animals, seems to have certain well-defined motives or functions: the search for food, for shelter, for a mate, and for amusement. Humans, whether they spend their days growing crops in the field or earning a salary in an office, move around for one or more of these reasons. So does a grazing cow, a displaying peacock or a playing puppy. Play, shown by many young mammals, is an interesting kind of movement, spontaneous and untaught. Humans and some domestic pets are among the few species that continue to play in adult life, and a large amount of activity has this motive, which may indeed be incompatible with the other motives for movement. The strenuous activity of amateur footballers in a match or mountain-climbers on a snow-slope is being performed not to seek food or shelter but purely for fun.

The type and speed of bodily movement are characteristic of the species. Cows and other herd animals walk slowly over a field grazing as they go, and even when driven, they rarely run. Rabbits graze and browse, moving only a few steps at a time, but can show a great turn of speed if suddenly startled. Dogs and horses can trot steadily, perhaps for hours. The swallow is the speed champion among birds and is said to attain a speed of 80 km per hour (50 m.p.h.), though it is not clear whether this is in a downward swoop, when gravity would assist muscular action. The condor, which can fly the length of the Andes, is undoubtedly a bird of great muscular strength, but the energy output of this feat would be difficult to calculate, since considerable use is made of soaring on air currents. The two extremes of types of running, sprinting and long-distance running, are displayed by the cheetah and the wolf in the animal world, and are described in the chapter on exercise; there are numerous varieties of animal between these extremes. These examples show the irregularity of performance of external work among species and among individuals. Even for two humans of the same age, sex and weight carrying out the same tasks in the same environment throughout the course of a day, one may have a far greater 'work-output' than the other. This is because people differ in their so-called 'unconscious work-output', their trivial, slight unconscious movements, or 'fidgeting', which are without motive, yet in sum make a greater or lesser demand on the energy supply.

ENERGY STORAGE: GROWTH AND BODY COMPOSITION

Energy taken in over a period of time must either be given out as heat loss or stored within the body, normally as additional tissue. Growing and pregnant animals are forming additional tissue and increasing in weight. The added mass during growth and pregnancy comprises the energy-rich components fat and protein, together with a small amount of carbohydrate; water, which is about two-thirds by weight of most tissues; and the mineral content, mainly bone. Studies of energy balance and nitrogen balance, which involve measurement of intake, excreta and heat loss for periods of anything from one day (in man) to several weeks or even months (in other animals), permit the calculation of total energy retained, and total protein retained, in the course of growth. The difference between total energy and protein energy is usually reckoned as fat energy, the small amount of carbohydrate being neglected.

The following example is typical of a one day balance for an adolescent boy, age 13 years, weight 62 kg:
Metabolizable energy intake = **12 800 kJ** = A
(found as described on p. 102)
Energy output (heat loss + work) = **12 100 kJ** = B
∴ Energy retained = $A - B$ = **700 kJ** = C
Protein intake = 60 g = 9.6 g nitrogen (protein is 16% N)
Nitrogen output = 8 g
∴ Nitrogen retained = 1.6 g = 10 g protein
Energy retained as protein = (10×24) = **240 kJ** = D (protein energy is 24 kJ per g)
Energy retained as fat = $C - D$ = **460 kJ**

It is during the periods of pre-natal life, infancy, and adolescence, that energy storage as growth is particularly striking, and observers have often commented on the large growth spurt, and huge appetite, of children at about the time of puberty.

For species other than man, it is of great biological advantage for the adult, as well as the young growing animal, to have a method of storing energy; such storage is particularly important for terrestrial species. Nearly everywhere on the earth's land surface there are seasonal variations in the availability of food, determined by cold or drought. So for any species whose life cycle is longer than a year, there will be times of plenty and of famine. Many species (including man) collect stores of food to be used later. Almost all species are capable of building an internal store of energy, in the form of fat.

Fat, having a high energy content per g, is a suitable form of energy store. Also it is light and has a low water content, so carrying it around does not waste much energy. Plants show an interesting contrast with animals, in the form of their storage energy. For the individual overwintering plant, energy is usually stored as starch or inulin (forms of carbohydrate) in roots, tubers or bulbs. In plants fat occurs mainly as oil in seeds and fruits, those parts which are passively moved away from the parent plant, often by wind or animals; so small weight may be an advantage.

All animals including invertebrates are capable of manufacturing and storing fat at all times of life, but certain species, especially when young, show a prodigious capability for this. A piglet during the first two weeks after birth can increase the proportion of fat in the body from 2% to 12% of its body weight. Since during this time the whole body weight is rising, this percentage increase may represent a deposition of 450 grams of fat, nearly half a kilogram, in two weeks. The young growing animal is also depositing protein quite efficiently, an ability which continues later in life although at somewhat reduced efficiency. This point is illustrated in Fig. 8.2, which shows a comparison of energy retention, for each increment of metabolizable energy intake, in pigs weighing 7 kg (about five weeks old) and pigs weighing 35 kg (about 12 weeks old). At the upper end of the scale of energy intake, from $1000 \, \text{kJ} \, \text{kg}^{-0.75} \, \text{day}^{-1}$ upwards (an abundant food supply), the lines for *total* energy deposited are very similar in the two age-groups, but in the amount of energy deposited as *protein* the young animals are well above the older; the younger pigs are laying down more protein and less fat for each increment of energy intake than are the older pigs. At the lower end of the range of energy intake (between 400 and $900 \, \text{kJ} \, \text{kg}^{-0.75} \, \text{day}^{-1}$) the difference in distribution of deposition of total energy and protein is in the opposite direction. At this 'under-fed' level of intake, the younger animals, and to a lesser extent the older ones also, are able to continue to deposit energy as protein, even when in overall negative energy balance; the lowest circles are below the zero-line on the ordinate scale which represents energy equilibrium (no growth). This must mean that even while depositing protein the pigs are losing fat. It is clear from observations of this kind that the body composition of farm animals can be altered by the level of feeding, and that the pattern of distribution of the stored energy as fat and protein may be very different in younger and older animals. The practical problem for animal husbandry is to attain in the

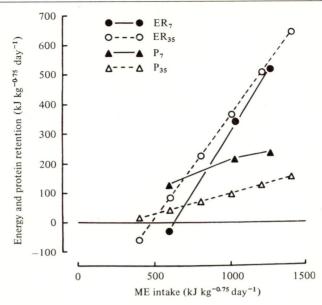

Fig. 8.2. Daily rate of deposition of total energy (circles) and of protein energy (triangles) by growing pigs, at various levels of food intake (increasing left to right along abscissa scale). Black symbols: young pigs about 7 kg body weight. Open symbols: older pigs about 35 kg body weight. ER = total energy retention: P = protein energy retention. By courtesy of Dr W.H. Close.

shortest possible time the particular distribution of protein and fat (and the total weight) in the bodies of pigs, sheep, poultry and beef-cattle which is desired by the consumer.

The point was made in Chapter 2 that by abundant feeding it is possible to lower an animal's critical temperature, that is, the ambient temperature at which its core temperature can be maintained only by an increase in metabolic rate. This point was illustrated in Fig. 2.7 which shows that a sheep's critical temperature was lowered from 30 °C to 20 °C by a high food ration. This was an adult animal, but the same is true for young growing animals. It is important for those who raise livestock for meat to make sure that the animal's metabolism is directed towards growth (i.e. increase in weight, deposition of fat and protein) and not diverted towards maintaining body temperature; this would be a waste of its food intake. So for the best productivity the animal should be kept at an ambient temperature above its critical temperature but not so high as to cause a decrease in its appetite. However the critical temperature itself varies with food intake and also with the animal's age. In consequence much effort has gone into defining the interaction of food level and ambient temperature in growth and development of farm animals. It appears for example that in pigs the storage of protein in the body is largely independent of ambient temperature and depends mainly on the quantity and quality of the food; the storage of fat, on the other hand, is dependent on both ambient temperature and food level.

Besides the young growing piglet, described on p. 105, two further examples of the rapid storage of a large amount of energy may be considered. One example is that of hibernating mammals. In autumn, squirrels may gain as much as 48 g in weight, one quarter of their total body weight. By the end of winter they have lost 40% of their autumn body weight. Another example is provided by migratory birds. Such birds eat enormously before migration. In a particular species known to migrate across the Gulf of Mexico from North to Central America, the 'fat index' (g fat : g non-fat dry matter) of the body was found to be three or four, before migration, and between 0.3 and 0.6 after migration, showing that a large amount of stored fat had been used during flight. (Typical fat-index figures for the bodies of non-migratory birds are 0.2 to 0.4.) The fat stored in the body of the migrants must of course have been used to make good the heat loss as well as supplying energy for the work of flying. It was calculated for this migratory species that the birds must have made a non-stop flight at an average speed of 40 km per hour (25 m.p.h.) for 60 hours. But again because of possible help by a following wind, the actual energy of the work-output is impossible to calculate.

MEASUREMENT OF ENERGY RETENTION

Comparative slaughter

The energy content or fat content of the animals described in the previous section was measured by direct analysis: that is, by bomb calorimetry of a sample of the dried carcass, and by weighing fat extracted by fat solvents from the dried carcass. For piglets, for example, one can thus compare the fat content of an animal slaughtered at two weeks with that of its litter-mate slaughtered at birth. Assuming that the two-week-old piglet had at birth the same percentage composition as its litter-mate, the gain of energy or fat in the two-week interval can be calculated. For migrating birds, carcasses of those picked up after being accidentally killed just before the flight can be compared with those picked up at the end of the flight. This technique is called 'comparative slaughter'. Clearly, a problem arises if one wants to record deposition of fat or of total energy in the same individual animal (or human) over a period of time, for example to find an effect of amount or composition of food or of environmental temperature on the rate of energy retention. For studies of deposition in individual animals, and for all studies on living humans, whether of energy content or of energy retention, some other technique must be used.

Methods for man and living animals

For fairly short-term experiments (a few days in man, several weeks in animals), it is possible to make precise measurements of *metabolizable energy intake*, by bomb calorimetry of samples of food and excreta, and of *energy output*, by oxygen consumption. The difference between the two values must be energy retained in or lost from the body. If one is simultaneously measuring nitrogen intake and output, and makes the assumption that all retained nitrogen is used for protein synthesis, one can calculate the amount of energy retained as protein. In the short-term, very little energy would be retained as carbohydrate, and most experimenters assume that the difference between total energy deposition and protein represents fat deposition.

To measure energy changes over a long period (months or years), it is necessary to determine the animal's or human's energy content at the start of the period, and again at the end, and so to find the deposition or loss by difference. Energy content of the living animal or human is found by measuring total weight of fat-free tissue, or total weight of fat, or both. If only one of the two components is measured, the other is calculated by difference from total body weight.

Fat-free tissue is measured by a dilution method, using the assumptions (a) that all the body water (or body potassium) is contained in fat-free tissue and (b) that water (or potassium) forms a constant percentage by weight of the tissue. If a precisely known amount of tritiated water HTO (a weak beta-emitter) is administered to the subject – animal or human – and allowed to equilibrate with the total body water, the concentration of beta emission, easily measured by scintillation counting, will give the volume of body water and hence the weight of fat-free tissue. If potassium is to be used, it is not necessary to administer a radio-active form because a radioactive isotope – $*K^{40}$, a gamma-emitter – occurs naturally and has a long half-life. It accompanies the stable isotopes K^{39} and K^{41} forming a constant percentage (0.011%) of the total potassium of all compounds. So a total body count of $*K^{40}$ allows calculation of the body's potassium and thus of the fat-free tissue which is assumed to contain 0.168% K^+ in men and 0.136% K^+ in women.

Total fat can be obtained by densitometry or – less accurately but more conveniently – by skin-fold thickness. Human fat has a density of 0.9. Fat-free tissue has a density of 1.1. The density of the whole body is obtained from the sum of the density of its two components.

So, if the density of the whole body is D, and there is $x\%$ of fat in the body, $\dfrac{100}{D} = \dfrac{(100 - x)}{1.1} + \dfrac{x}{0.9}$.

$$\text{(fat-free} \qquad \text{(fat)}$$
$$\text{tissue)}$$

So $x = (495/D) - 450$, where D is mass/volume. To measure volume, one uses the displacement or buoyancy method, weighing first in air and then in water (Fig. 8.3). The procedure requires various precautions and corrections. For an animal's carcass, air trapped in the fur or pelt could make the buoyancy erroneously high, so fur must be removed before the weighings. For man and other living animals, the air in the lungs gives additional buoyancy; in humans, this can be measured separately and allowed for in the calculation.

The measurement of total fat by skin-fold thickness involves measurement, by a specially designed pair of callipers, of a skin-fold at four particular sites on the body surface. The blades of the callipers are used to pinch a fold of the skin with its subcutaneous fat, with a precise pressure, and the distance between the blades is read off on a built-in scale. The greater the subcutaneous fat-layer, the thicker the fold. This method has been validated by comparison with the density method: the logarithm of the sum of the skin-fold thickness at the four sites bears a direct relationship to total body fat, as illustrated in Fig. 8.4. The relationship is different for different age groups and for the two sexes, and the

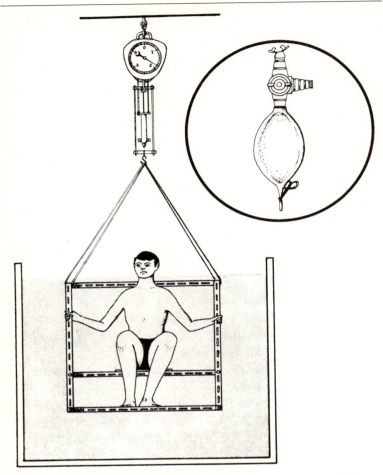

Fig. 8.3. Diagrammatic representation of underwater weighing apparatus. The inset shows the apparatus for measurement of residual air volume in lungs. From Durnin and Rahaman (1967).

method is subject to the error that the subcutaneous fat may not be a constant proportion of the total body fat. Nevertheless, approximate figures for total weight of body fat can be calculated from published data making use of the sum of the four skin-fold measurements.

When the weight of fat-free tissue and body fat have been found,

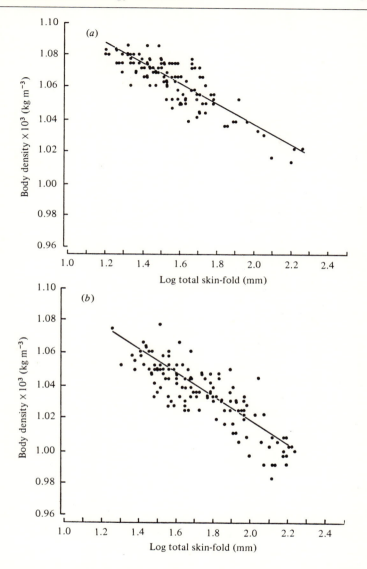

Fig. 8.4. Individual values for body density and the log sum of four skin-folds. (a): 116 men aged 17–29 years; (b): 129 women aged 16–29 years. See text for details. From Durnin and Womersley (1974).

total energy content can be calculated from the known kJ content
per gram of these two types of tissue.

CONTROL OF ENERGY INTAKE

The retention of energy in the body and the distribution of energy
reserves are influenced in the short-term by a number of hormones,
including the thyroid hormones, pituitary and adreno-cortical hormones, and adrenaline. In the long-term, however, the overall energy storage must depend on the relation of intake and output, and
it is the control of these which influences the storage of energy over
periods of weeks, months or years.

Obviously energy intake is controlled by the subjective feelings of
appetite and satiety. Humans certainly experience these feelings,
and it is presumed, by observing their behaviour, that other creatures do so as well. But appetite and feeding behaviour are not easily
explained in physiological terms. Herbivores and other vegetarians
among mammals and birds have a diet so dilute in energy that they
have to spend most of their waking hours eating or searching for
food. Carnivores and omnivores (lion, cat, pig, man) eat at intervals. Lions in the wild kill and eat every three or four days. Some
humans feel the need to eat every two hours; but when a group of
about five men of the Karamajong tribe in Uganda eat a cow,
leaving only horns and hooves, they will eat nothing more for four
or five days. Humans obviously differ greatly among themselves in
their feeding habits, and although the amount taken each day must
bear some relation to the work-output, this relation is not very
precise, and has a marked time-lag. A study carried out on army
cadets in training some years ago illustrates this point. Each cadet's
daily output of work and spontaneous food intake was measured
each day for 12 days; as Fig. 8.5 shows, the food intake ran more or
less parallel with the work-output, not of that day, or the day before, but of the day before that. An even more striking example of
the time-lag of food intake was provided by German prisoners after
returning home from Russian prisoner-of-war camps after the
Second World War. These ex-prisoners, on returning to an adequate
food supply after being on limited rations for many weeks, were able
to eat 29 300 kJ (7000 kcals) a day, for weeks and months, in the
course of restoring lost body weight – a capacity or 'appetite' about
twice that of a normal man during moderate work.

A series of experiments on rats and dogs in the 1950s and 1960s

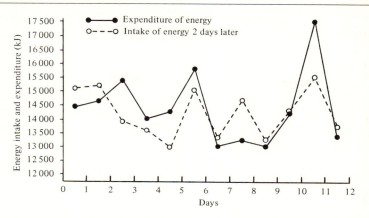

Fig. 8.5. Relationship between daily energy expenditure and intake 2 days later, in young military cadets. From Edholm, Fletcher, Widdowson and McCance (1955).

located a group of nerve cells in a specific region (ventromedial nucleus) of the hypothalamus, electrolytic destruction of which had a gross effect on the animals' food intake. A rat of 200 g which had been subjected to hypothalamic puncture of this small well-defined region would eat enormously so that within about three weeks it had become 400 g, and huge amounts of fat were deposited. It would then stop this excessive eating, and settle down to a steady food intake no greater than its spontaneous intake before the hypothalamic puncture, thus maintaining itself at 400 g. An electrolytic lesion elsewhere in the lateral hypothalamus caused the converse effect: the animal failed to eat and if not force-fed would eventually die. Experiments of this kind led to the postulation of a 'satiety centre' and an 'appetite centre' in the hypothalamus. The rat which overate did so because its 'satiety centre' had been destroyed, so it was unable to recognize that it had eaten enough. If such an animal, when 400 g, is then starved by being offered only a small amount of food each day, its body weight can be reduced to 200 g; if it is then offered food *ad libitum*, it will again overeat, rapidly return to 400 g body weight, and then maintain this weight. This observation raises several problems. If a 'satiety centre' has been destroyed, why does the rat not go on overeating indefinitely? Is the function of the hypothalamic centre to act as a 'lipostat', sensing in some way the total amount of fat in the body? If this were so, one might suppose that its destruction had resulted in setting the 'lipostat' at a higher level. Again, what would be the signal to the hypothalamus? Could it be the free fatty acids, or some other metabolite produced by fat depots, the plasma level of which might signal to the hypothalamus the total amount of turnover rate of fat depots? Although there is general agreement that hypothalamic 'centres' or 'neurone pools' are involved in the control of food intake, the sequence of the mechanisms involved is not yet clear. A short-term control, the feelings of hunger and satiety before and after each meal, may involve such transient sensations as the feeling of fullness in the stomach. One idea, the 'glycostatic theory', is that the intake is controlled by the rate of entry of glucose into cells in the ventromedial nucleus of the hypothalamus; if plasma glucose falls too low, the rate of entry falls, and this is the signal for eating. Another idea, the 'thermostat theory', related the sensation of eating at the end of a meal to the slight warming of the blood caused by the food's specific dynamic action. These might explain immediate short-term regulation associated with meal times, but not such a phenomenon as the enormous and sustained appetite of the returned prisoners-of-war. The fact remains that by some means or other, animals with an adequate food supply manage to control energy intake in such a way as to maintain a fairly constant body weight throughout their adult life.

OBESITY

The excessive storage of energy leads to obesity, a condition in which the amount of adipose tissue is inconveniently great, causing discomfort and perhaps even a strain on the cardiovascular system or heart muscle in carrying the excess weight. It occurs because at some time in the person's life his energy intake has been in excess of his current output. The daily excess may have been quite small but the effect is cumulative. A person who eats 400 kJ (100 kcals) too much each day (the energy content of one thick slice of bread), would over 15 years have eaten 2200 MJ too much. If this had all been stored as adipose tissue, it would have made 56 kg fat. Even for a man maintaining a perfect balance at age 20, this same intake at 50 would be excessive because as one ages one's resting metabolic rate is decreased (Chapter 2). In the age range 14–20 years, a man's metabolic rate is about 320 kJ per square metre per hour ($kJ\ m^{-2}\ h^{-1}$); in the age range 50–60 years, it is 257 $kJ\ m^{-2}\ h^{-1}$. So between age 20 and age 50, there could have been a decrease of 1200 kJ per day in resting metabolic *requirement*, an amount of energy contained in about three slices of bread. So his energy intake should have gone down by this amount over the 30-year period if he is to remain in balance. This is on the assumption that his muscular work-output is the same at 50 as it was at 20. If, as is often the case, it is much less in the older person, his energy intake should have decreased even more, if he is to remain in energy equilibrium. In practice, people tend to eat regular meals of constant size by force of habit, and as the years pass this food intake gradually becomes further and further in excess of their requirement for resting and active metabolism. This is the commonest cause of obesity: the failure to match one's food intake to one's decreasing energy requirement and output over the years. A person may become aware at say age 50, of being overweight, and therefore reduce his food intake, but this may merely keep him in equilibrium at that already excessive weight and do nothing to reduce it.

The treatment of obesity is quite difficult. A possible method is to take frequent small meals of mainly protein foods. The heat increment of feeding (see Chapter 2) is greater for protein than for carbohydrate or fat, so after a protein meal there is a large production and loss of heat; this transient increase in metabolic rate helps towards the balancing of intake and output. Fig. 8.6 shows these effects, in a person whose metabolism was measured twice in a whole body calorimeter for a 28-hour period, first taking a high carbohydrate diet and then, on another occasion, a high protein diet. The difference in effect on metabolic rate of these two diets was very obvious just after the meal, and during spurts of exercise on a bicycle ergometer. To make the best use of the heat increment effect of protein, during weight reduction, the meals must be frequent, but of course small in total kilojoules.

One problem encountered in reducing a person's total kilojoule intake when treating obesity is that his resting metabolic rate may go down as his intake decreases. As explained in Chapter 2, heat production varies with food intake. A fall of daily intake from a 'medium' intake, 8900 kJ (2127 kcals) to a 'low' intake, 4400 kJ (1052 kcals) may be accompanied by a 6% reduction of resting metabolism. If energy output as well as intake has decreased, the excess of intake may still be considerable. Fig. 8.7 illustrates the energy output of one individual during a 28-hour period, on three different occasions, on a 'high', 'medium' and 'low' energy intake. It

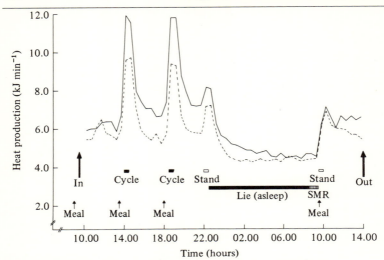

Fig. 8.6. Metabolic rate measured continuously in a whole body calorimeter, on the same subject, during each of two 28-hour periods; in one period he was on a high carbohydrate diet (dotted line) and in the other, a high protein diet (continuous line). SMR: period after waking during which standard metabolic rate was measured. From Dauncey and Bingham (1983).

shows how much the metabolic rate, and thus the body's requirement for energy, may vary with the intake. Obviously, to make a low energy intake effective in treating obesity, it must be accompanied by a deliberate raising of the requirement for energy, by taking additional exercise. This would help to offset the effect of the fall of resting metabolism consequent on the fall of food intake.

SUMMARY

The adults of most species of homeotherms can maintain energy equilibrium: their overall energy intake over a period matches their output. The young growing animal and the pregnant female are in positive energy balance. Intake and output are measurable with a fair degree of precision, but the amount of energy stored, mainly as adipose tissue, is not easy to measure in the living animal. The amount and distribution, as between fat and protein, of energy stored by young growing animals of the livestock species is of some economic importance. Appetite and satiety in all species, and in man

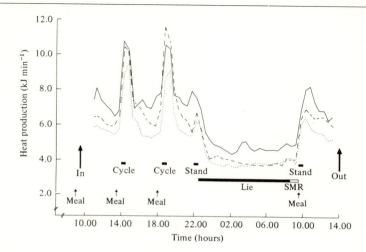

Fig. 8.7. Metabolic rate of male subject aged 48 years, weighing 55.9 kg, during three sessions of 28 h in the whole body calorimeter. Energy intakes were high, 14 736 kJ per day (———); medium, 8895 kJ per day (- - - -); or low, 4444 kJ per day (····), and of identical contributions from fat, carbohydrate and protein. Cycling was at the low work load of 0.5 N at 50 revs per minute. Mean values of heat production are plotted. Note that the metabolic rate is least on the low energy intake. SMR as for Fig. 8.6. This figure was used in Chapter Two to illustrate the effect of meals and work on heat production. From Dauncey (1980).

habit and social behaviour as well, determine energy intake; the physiological control is centred in the hypothalamus but the mechanism is still uncertain. Stored energy in man is sometimes excessive, leading to obesity which, once established, is difficult to cure.

CHAPTER NINE

Muscular exercise and its limitation

PHYSIOLOGICAL ADJUSTMENTS TO EXERCISE

Even in basal conditions when a man is lying at rest in a post-absorptive state with all his limbs as fully relaxed as possible, some muscular work is still going on. His heart is beating, his respiratory muscles are contracting rhythmically, his gut muscles are continuing peristalsis; even his apparently relaxed limb muscles are showing tonus, the continuous slight asynchronous activity of groups of muscle fibres which takes place throughout life. When the man stands up and starts to run, there is a sudden enormous increase in the energy output of his skeletal and cardiac muscles. The cardiovascular and respiratory adjustments which take place at the beginning of such physical exercise, which have the effect of supplying additional nutrient and oxygen to the active muscles, are well-known as a matter of common experience: in brief, the heart beats more quickly, the person starts to breathe more quickly and perhaps more deeply also, and soon feels warm. The mechanisms of the various cardiovascular adjustments to muscular exercise are now fairly well explained; the respiratory adjustments, in spite of much study, are still not fully explicable.

A study of the precise sequence of events at the beginning of strenuous exercise (such as running a race or running to catch a bus) helps to explain their physiological mechanism. These events, well described in text-books of physiology, will be summarized only briefly here. The initial event, which may indeed occur even before the running starts, is an anticipatory increase in the activity of the whole sympathetic nervous system, in consequence of the mental excitement. The so-called 'higher centres' of the brain, whether the brain of the athlete starting a race, of the traveller in danger of missing the bus, of a startled rabbit, or of the gazelle escaping from a leopard, initiate a burst of activity in all sympathetic nerves. These nerves in turn cause an increase in heart rate (chronotropic effect) and in stroke volume (inotropic effect), thus raising cardiac output.

They constrict arterioles in skin and abdominal organs, and some sympathetic vasodilator fibres may also dilate blood vessels of skeletal muscle: this elicits a massive redistribution of the blood volume towards the skeletal muscle and thorax. Two further minor effects of this initial burst of sympathetic activity are (1) to stimulate release of adrenaline from the adrenal medulla; this in turn causes a rapid rise of blood glucose by enhancing liver glycogenolysis; and (2) to dilate the bronchioles, thus lowering resistance of the airways of the respiratory tree.

All these sympathetic effects may take place even before the start of the running and certainly within a few seconds after the start. They occur, too, if the man or animal simply stands still, feeling excited or frightened, but not actually running. In this case, of course, the metabolic rate and oxygen consumption rise transiently and soon decrease again. If, however, the running does start, a further group of events and adjustments begins.

One of these is the activation of sensory receptors in the main joints and muscles, especially of the legs; these are the tendon organs and muscle-spindles which send a burst of impulses to the sensory cortex of the brain, which in turn activates the respiratory centre to give an increase in the rate and depth of breathing. Another event is the squeezing of the veins of the legs by the rhythmically contracting and relaxing leg muscles: this speeds up the movement of venous blood back towards the heart, and this increase in the rate of venous return to the already rapidly contracting heart greatly increases cardiac output. The metabolism of the muscle fibres themselves is, right from the start, at least partly anaerobic, and such metabolic processes release into the muscle capillaries various 'metabolites' (lactic and pyruvic acids, K^+ ions and adenosine diphosphate have been implicated). These have the effect of dilating the muscle capillaries, thus increasing muscle blood flow and bringing the blood into close contact with the actively contracting muscle fibres. After a few minutes, or even seconds, the heat generated in the contracting muscles warms the whole circulating blood enough to cause warming of the hypothalamic region of the brain, which in turn produces temperature-lowering responses: skin blood vessels now start to dilate, and (in man) sweat is secreted. At this stage, too, there is a sufficient build-up of lactic acid in the circulation to stimulate chemoreceptor cells of the carotid and aortic bodies, neural impulses from which impinge, directly or indirectly, on the respiratory centre and cause continuation and enhancement of the over-

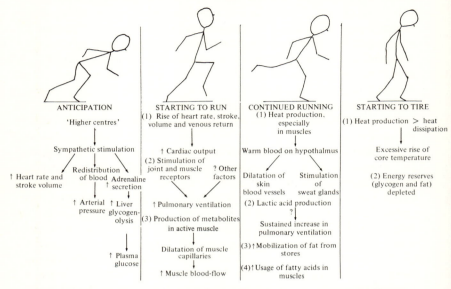

Fig. 9.1. Summary of some of the physiological changes occurring during running in humans. Long arrows pointing downward mean '*causes*'; short arrows pointing upward mean '*rise of*'.

breathing. Apart from the increased flow through the capillaries of the working muscles, all these responses are readily observable. The individual is aware, when running or taking any strenuous exercise, of his fast and deep breathing, quick and strong heartbeats, dilated veins, and flushed, warm and moist skin. In sum, changes in respiration and in the movement and distribution of blood have the effect of greatly increasing the glucose and oxygen supply to the active muscles, removing from them the products of metabolism, removing carbon dioxide, and dissipating excess heat from the body. An outline of the physiological adjustments at various stages of a run is shown in Fig. 9.1.

UNEXPLAINED RESPONSES

Although the mechanism of many of the responses to exercise can be fairly well explained, several mysteries remain. One is the precision with which oxygen intake matches the exact level of work during sustained exercise. When a man running steadily at 8 k.p.h. suddenly starts to run at a steady 12 k.p.h., within seconds his respiratory rate or depth has adjusted to the new level of work. It is difficult to explain this in terms of additional activity of joint receptors or a precise level of blood lactate. It is also difficult to explain, in terms of stimulation of joint receptors, any respiratory effect of isometric work (such as tug-of-war or weight-lifting) in which the joints may be locked in one position for seconds or minutes. Another incompletely explained phenomenon is the dilatation of capillaries in the active muscle, by what are loosely termed 'metabolites'. In experimental conditions several substances known to be released by anaerobically-contracting muscles dilate vessels and increase blood-flow when perfused through the circulation of an inactive muscle. Which of these, if any is the effective substance in normal conditions? Or may all of them be working together to enhance one another's effect? It is known, at least, that the greatly increased muscle blood-flow is not simply the effect of deprivation of oxygen (hypoxia) as such. In one experiment on the hind-limb of a dog, the rate of blood-flow to the limb, and the oxygen tension (P_{O_2}) of the arterial inflow and venous outflow to and from the limb, were continuously recorded. As soon as the limb started to move, blood-flow increased and venous P_{O_2} fell to almost zero, the arterial P_{O_2} remaining unchanged; thus the arterio-venous oxygen difference was enormously increased. Steady levels, high and low respectively, of blood-flow and venous P_{O_2} were maintained during the steady exercise of the limb. Then, as soon as the exercise stopped, venous P_{O_2} instantly rose to its pre-exercise level, whereas the hind-limb blood-flow only slowly fell back to the pre-exercise level. If hypoxia itself had been the cause of the enhanced blood-flow, the flow would have been expected to return to the resting level at once, when the oxygen supply would be adequate for the now reduced metabolic requirement of the muscles. However, 'metabolites' accumulating in the muscles might take some minutes to be washed away completely, and during this washing-out process the residues would have continued their blood-flow-enhancing action. It may therefore be surmised that it is the presence of one or more chemicals, rather than the absence or deficiency of oxygen, which causes the enormously increased muscle blood-flow, a phenomenon of great physiological importance in exercise.

THE LIMITATION OF EXERCISE

It is clear from this outline of the body's response to exercise that, apart from the initial burst of sympathetic activity associated with mental excitement or fear, all the responses are produced by the working of the muscles themselves. The pumping of venous blood back to the heart, the stimulation of joint receptors, the build-up of circulating metabolites, and the warming of the blood occur because the muscles are moving; these processes in turn lead to the increased supply of nutrient and oxygen, the removal of carbon dioxide, and the prevention of excessive rise of body temperature. It is, in fact, the direct mechanical, physical and chemical effects of the muscles' own activities which produce those adjustments of the cardio-vascular and respiratory systems which, in turn, make possible the continuance of the muscular action. Why, therefore, cannot the exercise continue indefinitely? All these adjustments of the body to exercise seem so complete and satisfactory that one may feel surprised that the runner eventually has to stop running. What is it that limits the performance of muscular exercise, forcing the jogger, runner or swimmer to his break-point, and even the trained athlete, on occasion, to a state of collapse?

THE BREAK-POINT

In seeking for an explanation of the break-point, the point at which the exercising human or animal is forced to stop, several possible factors may be considered: (1) the oxygen intake could be inadequate either because of insufficient effort of the respiratory muscles, or because of too slow a rate of exchange at the alveolar surface of the lungs; (2) the cardiac output, though enormously higher than in the resting state, could be insufficient to deliver the oxygenated blood to the working muscles; (3) there could be local inadequacies in the blood supply perhaps to the active muscles themselves; (4) the break-point might represent the point at which the supply of fuel (carbohydrate and fatty acids) to the exercising muscles has become exhausted; (5) the rise in deep body temperature which always occurs in strenuous exercise in spite of the various heat-dissipating mechanisms may in some way limit performance when a certain temperature is reached; or (6) some other as yet unknown factor, or a combination of factors, may be operating. It must be said at once that no single explanation of the break-point, applicable to all situations, has been found. However, each of the factors listed above will be considered.

Long-continued exercise always causes panting, and a certain amount of discomfort in breathing (dyspnoea), and one may have the subjective feeling of being unable to continue because of 'breath-lessness'. It has been found, however, that exercise is not limited by the power of the respiratory muscles to move air in and out of the lungs, nor by the rate of gas exchange at the lung surface. In both untrained men and trained athletes, respiratory movements at the break-point are in fact *greater* than necessary to keep the blood saturated with oxygen; and in severe exercise the arterial blood leaving the heart may indeed have a P_{O_2} slightly higher (and P_{CO_2} slightly lower) than at rest. Clearly, performance is not limited by ability to oxygenate the blood or to remove CO_2 from the body.

The suggestion that performance is limited by failure on the part of the cardio-vascular system, the distribution of oxygen rather than its intake, is a plausible one. Undoubtedly, a large part of the muscles' energy during sustained work is due to anaerobic metabo-lism, as shown by accumulation of lactic and pyruvic acids in circu-lating blood. If, during sustained exercise, one breathes an artificial gas mixture that contains 66% oxygen (instead of atmospheric air which is only 20% oxygen), one's ventilation rate is slower, but one can sustain a given work-load for a longer time than during air-breathing. It would seem that the additional partial pressure of oxygen in the tissues decreases the usage by the muscles of a largely anaerobic metabolism. If, however, the break-point is indeed the point at which there is a failure to deliver oxygen to the muscles at an adequate rate, several further questions arise. Is it inadequacy of cardiac output, or of muscle blood-flow, or of total circulating blood volume, in view of the fact that sustained exercise requires a rapid flow through the skin as well as through the working muscles?

Biochemical evidence supports the suggestion that, at least in long-distance running, the fuel supply of the striated muscle may be an important factor in limiting performance. It has been calculated that the amount of a man's muscle glycogen, together with glucose which the liver could supply by glycogen breakdown, would last for only about 60 minutes, if these were the only source of fuel for active muscles. Glycogen, however, is not the sole source of fuel: right from the start, part of the fuel supply to the active muscles consists of free fatty acids brought by the blood from adipose tissue, and the fuel consisting of this mixture of carbohydrate and fat gives maxi-mum power to striated muscle. When the carbohydrate supply runs out so that the muscle is working on fatty acids only, the power of its contraction at once decreases. So the Marathon runner should finish the race just at the point where he has no more available muscle glycogen in his leg muscles. If some had remained, then he could have run faster; if it had become exhausted before the finish, the final stage of the race would have been run at less than maximal power, on fatty acids only. The idea that availability of carbohydrate may indeed be a factor in limiting performance is supported by the effectiveness of small doses of glucose in helping those who collapse in exhaustion after strenuous effort. However glucose should not be taken immediately *before* a long distance run, because the consequent rise of blood glucose would stimulate insulin secretion, and one of the effects of this hormone is to decrease lipolysis in adipose tissue, thereby limiting the muscles' supply of free fatty acids at precisely the wrong moment.

Information about the mechanism of the break-point may be gained by further study of the symptoms of athletes who collapse at the end of a race. They suffer from violent cramps of the muscles, a pain known to be associated with the accumulation of lactic acid, and the skin is noticeably pale and cold. This pallor suggests constriction of the vessels carrying blood to and through the skin. In consequence, heat loss is reduced, core temperature increases, and the athlete may show symptoms of heat-stroke. It is not clear what causes the increase in sympathetic vasoconstrictor tone to the skin blood vessels, but the consequence is a re-distribution of the blood volume away from the periphery of the body towards the heart and brain.

Perhaps, then, the limitation of exercise is due for some reason to a rise in core temperature. This seems a probable explanation since the performance of Marathon runners is likely to be better on a cool day, when a rapid heat loss from the skin surface is possible, than on a hot one, and an experiment on trained runners has quantified this common observation. Three trained athletes (middle-distance runners competing in the Commonwealth Games) were asked to run on a treadmill at the speed at which they normally trained, for 40 minutes, or longer if they felt able to do so. Each runner performed this run at two different dry-bulb temperatures: either 16 °C (61 °F) or 26 °C (80 °F). At the lower temperature, all three continued for at least 45 minutes, and felt that they could have gone on longer; their body temperatures increased by 2 °C, on average. At the higher temperature, all three managed the 40 minutes, but none could continue longer. Their body temperatures increased, on average, 4 °C; one runner finished with a core temperature of 40.4 °C (104.7 °F). Obviously, at the higher environmental temperatures, the runners could not lose heat fast enough; and this observation strongly suggests that, at least at high ambient temperatures, rate of heat loss may be the limiting factor at the break-point.

Actual heat-stroke involves heat damage to brain structures, and the brain temperature of running athletes may in fact be slightly below that of the rest of the body core. Even though man does not possess a rete mirabile of the kind well-identified in some other mammals, there is nevertheless countercurrent heat exchange between the arterial blood supply to the brain and the cooled venous drainage from the facial area from which evaporation occurs during exercise (Cabanac and Caputa, 1979). This apparently keeps brain temperature appreciably below that of the body core generally. Athletes often go into a state of what appears to be heat syncope (fainting) only after they have stopped running. It has been suggested that this is because with the cessation of air movement across the face, there is a reduction in the rate of evaporation and thus of the extent to which the blood passing through the facial skin is cooled; the brain temperature then rises abruptly towards that of the general core temperature.

The suggestion that a rise of deep body temperature causes cessation of exercise is supported by comparing the sprinters of the animal kingdom – the cheetah and the antelope – with the long-distance runners, wolf, caribou and wild dog. The cheetah and the antelope can achieve maximum speeds of 110 k.p.h. and 90 k.p.h. respectively (69 and 56 m.p.h.), but only for short periods. In the bush, a cheetah rarely runs a distance of more than 1 km. If it does not kill in this short sprint, it lies down in the shade for 5 or 6 hours, before making a further attempt. Of the heat generated in the sprint, 70% is stored in the body, and when the deep body temperature reaches 41 °C the animal lies down. (This resting, besides permitting body cooling, would also allow repayment of the oxygen debt incurred during the sprint, which is almost entirely an anaerobic action.) The wolf or wild dog, while running-down its prey perhaps for many kilometres, is losing heat from the body via the respiratory tract, and only 4% of the heat generated during the activity is stored in the body, so that the animal could continue for a long time before showing a dangerous rise in deep body temperature.

PHYSICAL TRAINING

Whatever the explanation of the break-point, there is no doubt that physical training greatly increases the amount of work which an

individual can do before he reaches his break-point, either by allowing exercise to continue at the same rate for a longer time, or by
allowing more intense exercise (such as faster running), or both. The
study of the physiological basis of muscle training is of interest to
professional trainers and athletes who are always trying to improve
their performance. Such a study is also of interest to physiologists
for the light it may throw on the performance of work, and its
limitation.

The effects of training can be assessed in several ways. For example, the amount of work done, or the rate of accumulation of blood
lactic acid, could be examined. Moreover, the chosen parameter can
be compared in the same individual before and after training; or the
performance of a whole group of trained athletes can be compared
with that of a group of untrained people of similar age, weight and
sex. An interesting comparison of the blood levels of some metabolites known to be concerned in a muscular exercise has been carried
out on a group of trained amateur athletes, members of a long-
distance running club, the 'Thames Valley Harriers', and a group of
men and women of similar age, quite fit but unaccustomed to this
sport (Corbett, Johnson, Krebs, Walton and Williamson, 1969). All
participants ran for $1\frac{1}{2}$ hours, the trained runners at the rate of
16 k.p.h. (10 m.p.h.) and the untrained people at the rate of
11 k.p.h. (6.9 m.p.h.). Blood samples were taken before, twice during, and for two hours after, the period of running. The level of
ketone bodies in the blood of the untrained people was found to rise
during the exercise and remained at a high plateau level for at least 2
hours afterwards; the level in the trained people was low and unal-
tered during and after the run (Fig. 9.2). Levels of free fatty acids
and of glycerol were raised in both groups but were higher in the
untrained group; glucose levels were the same in both groups. These
results suggest that one of the effects of training may be on the
biochemistry of the muscle: an improvement in the muscles' ability
to take up and metabolize fat.

An experiment designed to compare the effects of different
amounts of training was carried out on a group of young women
students at a physical training college (Knibs, 1971). Each student's
maximum work-output was first measured by finding her maximum
oxygen consumption during hard pedalling on a bicycle-ergometer.
Then, for six weeks, the students undertook 20-minute training sessions on the bicycle-ergometer at various intervals and rates of work,
as follows:

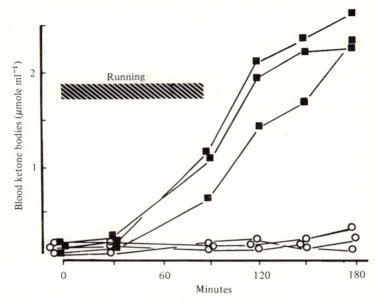

Fig. 9.2. Concentration of blood ketone bodies (acetoacetate and 3-hydroxybutyrate) in three athletes (○–○) and three non-athletes (■–■). Redrawn from Corbett, Johnson, Krebs, Walton and Williamson (1969).

1 session per week		3 sessions per week	
50% of maximum (Group 1)	80% of maximum (Group 2)	50% of maximum (Group 3)	80% of maximum (Group 4)

The results of the training were assessed by the effect on the blood lactate, and also by the effect on the work-output at a standard heart rate, 170 beats per minute, or ('W 170'). For Group 4, those having the most vigorous training, the results were:

	Work rate, 'W 170'	Blood lactate
	kg m min^{-1}	mg 100 ml^{-1}
Before training:	700	40
After training:	940	30

For Groups 2 and 3, the results were less, though in the same direction; the training of Group 1, once a week at 50% of their maximum, had no effect at all.

These experiments illustrate some of the ways in which one can

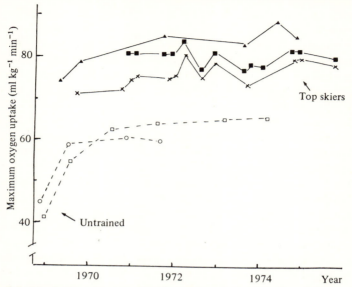

Fig. 9.3. Maximal oxygen uptake of cross-country skiers, and of two subjects who started intensive training in 1969. Redrawn from Åstrand and Rodahl (1977).

quantify the effects of physical training, effects familiar in a qualitative sense to anyone who participates in any form of sport. They do little to elucidate the precise mechanisms by which the body's energy metabolism changes during the process of training. Such mechanisms are the subject of research in laboratories in many parts of the world, and much of the pioneer work in the field was done in Scandinavia. Cross-country ski-racing, perhaps over steep terrain in intense cold, makes a tremendous demand on the participant's energy supply; obviously the physical training necessary for this sport must also be intense. Fig. 9.3 shows the skier's very high maximum oxygen intake during strenuous activity, a good index of athletic fitness. A number of specific effects of such training on the cardio-vascular system, blood, and muscle fibres have been defined. (1) The stroke volume of the heart increases. This is at least partly due to physiological hypertrophy of the cardiac muscle fibres, a process also occurring in any much-exercised striated muscle, and it would give a more vigorous heartbeat. (2) There is a greater uptake of oxygen per breath. This might be explained if, by training, all the lung alveoli were fully dilated all the time, instead of having groups of alveoli partially collapsed, the normal state of affairs in untrained persons. (3) The total blood volume increases, and, later, there is an increase in the amount of haemoglobin in the body but not its percentage in the blood. Presumably the blood volume rise must be a matter of transfer of intracellular or interstitial fluid into the vascular system, but its mechanism is hard to understand. The increase in plasma volume might indeed occur quite quickly: after a mere 47 km (29 mile) hill walk, the plasma volume (though not the red cell volume) of the walkers can be observed to have increased. (4) There is an increase in the number of muscle capillaries. (5) There is an increase in the concentration of myoglobin in the red muscle fibres. Myoglobin functions not only in short-term oxygen storage but also in speeding the diffusion of oxygen through cytoplasm to the mitochondria. (6) Several other changes occur in the structure and enzyme content of the muscle fibres, such that their oxidative metabolism is increased. There is an increase in the number and size of the mitochondria; and an increase in the concentration of the enzymes concerned in the oxidation of carbohydrate, fatty acids, ketones and pyruvate. These changes occur mainly in the 'slow twitch' fibres, also called Type IIA fibres. (Fast twitch – Type IIB or white – fibres have very few mitochondria and use glycolysis, not oxidation, as their energy source. Type I – slow red – use oxidation.) Since training thus increases the oxidative capacity of muscles, much fatty acid can be used as an energy source, as shown by a lower respiratory quotient. Muscle stores of glycogen are depleted more slowly so capacity for long-continued exercise improves. The general effect of training on the intensity and duration of the muscles' aerobic metabolism is indicated in Fig. 9.4, in which the aerobic metabolism of Marathon runners and non-athletes is compared. During submaximal exercise, the improved local oxygen transport in muscle – by more capillaries, myoglobin, mitochondria and enzymes – reduces the requirement of the exercising muscle for arterial blood, and after training the muscle cells can work at a lower tissue P_{O_2} because of their improved oxidative capacity. All these changes are reversible: the muscles and heart can be de-conditioned by bed-rest as quickly as they can be trained by exertion. The mechanisms bringing about these biochemical and anatomical changes during training and de-conditioning are still a mystery.

CONCLUSION AND SUMMARY

The mechanical behaviour of muscles and joints, and the biochemistry of actively contracting muscles, are thoroughly discussed in most

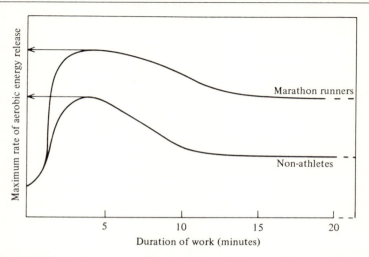

Fig. 9.4. Relation between endurance time and the rate of energy release from aerobic sources. Reproduced with permission from Bassey and Fentem (1981). Copyright: Academic Press, Inc., London.

text-books of physiology and biochemistry. Much experimental work has been done on isolated muscles or muscle groups, in defining, for example, the efficiency of the biceps at different rates of work, or the efficiency of running with different angles of the joints at ankle, knee and thigh. This present account of the metabolism of exercise has been concerned with the body as a whole, and especially with the cardio-vascular and respiratory adjustments involved in supplying energy in the form of free fatty acids and glucose to the actively working muscles, and the oxygen required for the oxidation of these substances. This chapter may seem to have raised more questions than it has answered. There is the continuing puzzle of why work-output and oxygen consumption of the whole body are so exactly proportional, even though the muscles' metabolism at high work-rate is largely anaerobic. The initial and continuing stimulus to over-breathing is not fully explicable. At the limits of exercise, it cannot yet be said which, if any, of the muscles' metabolites produce the cramps, and how they do so, and what causes the skin vasoconstriction at collapse. The mechanism of the increase in haemoglobin and myoglobin, in the number of muscle capillaries, and in mitochondria and oxidative enzymes as a result of physical training, are other problems which remain to be solved.

Fever

ALTERATION OF SET POINT TEMPERATURE

The hyperthermia accompanying muscular exercise is a normal state: heat production temporarily outstrips the rate of heat loss, and the core temperature is raised. It rapidly falls back to the resting level at the end of exercise. This chapter is concerned with an abnormal state of hyperthermia: raised body temperature associated with a state of disease. Fever can be defined as an elevation in body temperature to a level which is statistically above the range of the daily variation in body temperature of a resting subject and which is generally associable with a pathological condition. In fever, thermoregulatory activities operate to bring about the rise in body temperature. This clearly distinguishes fever from other hyperthermic conditions such as those induced by a high heat uptake from the environment or a high heat production during intense bodily activity, where the thermoregulatory activities are operating to resist the rise in body temperature.

The pattern of thermoregulatory activities during fever is as if the set point of deep body (or core) temperature has become displaced in an upward direction. Consistent with this interpretation, during a sustained steady-state fever, thermoregulation functions operate quite normally but to sustain a stable core temperature at the elevated level. The pyrogenic (fever-producing) agent somehow acts upon the central nerve cells involved in the regulation of core temperature in such a way that the defended stable level of core temperature is raised. The immediate effect of this is that the normal core temperature is interpreted as being hypothermic. Processes of heat production and heat conservation are activated, and cause core temperature to rise until it equals the febrile set point value. Thereafter, for so long as the fever persists at the level, the elevated set point is defended as if it were normal.

When the production and the action of the pyrogen ceases, the defended level of body temperature returns to its normal value. The effect of the withdrawal of the pyrogenic factors is as if core temperature is then interpreted as being above the now normal set point. Processes are activated by which heat production is decreased, and heat loss is increased. Core temperature then falls until it again equals the normal set point. Accordingly, during the developing phase of fever the skin is vasoconstricted, and the subject is shivering and huddled, maximizing heat production and conservation. During the recovery phase the skin is vasodilated, and the subject sweats and lies sprawled out, in such a way as to maximize heat loss. Fig. 10.1 illustrates this sequence of events diagrammatically.

This functional normality of a steady-state fever was clearly demonstrated in a study of the thermoregulatory responses of men to periods of exercise. During the study, one of the subjects developed a fever but insisted on completing the exercise regime. His thermoregulatory responses to exercise were found to be similar to those of the healthy subjects, and it was evident that his elevated core temperature was defended just as were those of his healthy colleagues. More readily repeatable, and confirmatory, observations have been made on other mammals in the laboratory.

The description of fever as the temporary elevation of the thermoregulatory set point poses the question: 'What causes the elevation of the set point and what is the mechanism by which the causative agent works?'

PYROGENS: BACTERIAL ENDOTOXIN AND ENDOGENOUS PYROGEN

Fevers are almost always the consequences of an infection by an invading bacterium, virus or protozoan. The fever results from the effect of a toxic product of the invading microorganism, a *bacterial* (or viral) *endotoxin*, acting on the phagocytes of the host, which respond by producing another substance, *endogenous* (or *leucocytic*) *pyrogen*, which in turn acts on the central nervous tissues concerned with thermoregulation. These two substances will now be described.

Pathogenic bacteria, particularly the gram-negative ones (that is, those taking up the counter-stain in Gram's staining method), have been shown to produce a water soluble heat-stable lipopolysaccharide (mol. wt $5 \times 10^5 - 2 \times 10^7$) which is pyrogenic (fever-producing) when introduced in minute quantities (0.01–0.001 μg) into mammals including man, dog, cat, rabbit and horse. This pyrogenic product of microorganisms is now generally referred to as a bacterial endo-

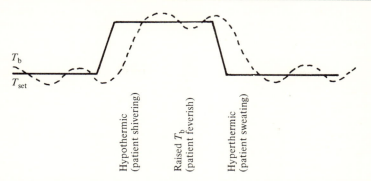

Fig. 10.1. Diagram showing the relation of set point temperature (T_{set}) and body temperature (T_b) before, during and after a period of fever.

toxin, rather than bacterial pyrogen, since it has many toxic influences quite apart from its pyrogenicity.

A pyrogenic substance can also be obtained by incubating (at 37 °C) polymorphonuclear phagocytes which accumulate in the peritoneal cavity of rabbits following the introduction of saline into the cavity, or by incubating leucocytes obtained from blood in the presence of a small amount of bacterial endotoxin (either as dead organisms or purified). This endogenous pyrogenic substance is a small molecule protein, which is denatured by heat. It is, therefore, chemically quite different from a bacterial endotoxin. True fever (the raised set point syndrome) may accompany infection by gram-positive organisms and by viruses as well as by gram-negative organisms, and may also occur in the absence of any evidence of infection. An acceptable account of the genesis of fever must allow for all these occurrences.

It is now established that the fevers which accompany infections by gram-negative and gram-positive organisms and by viruses all originate in the release of toxic products. These products are not identical either in their sites of production in microorganisms and viruses, or necessarily in the finer details of their effects on body tissues. There is, however, a general pattern in the relationship between these various toxins and the body's reactions.

The phagocytic cells of the reticulo-endothelial system (of which the polymorphonuclear leucocytes are a major constituent, but which also includes the monocytes, macrophages and the Kupffer cells in the liver) take up the circulating endotoxins as well as the

invading microorganisms. Evidently the phagocytes respond to the
endotoxins in several ways including the activation of antibody formation and the production and release of the endogenous pyrogen.
They constitute the body's primary defence.

There is also evidence that invaders other than microorganisms
elicit from phagocytes the production of endogenous pyrogen. The
fever associated with cancerous growths, for instance, involves the
production and release of endogenous pyrogen. Whether the cancer
fevers stimulate the phagocytes because endogenous products of
cancer cells are recognized as foreigners, or whether it is because
there is an exogenous (viral) cause for the genesis of the growths, is
not yet known.

It was thought at one time that the bacterial endotoxin itself acted
on the hypothalamus to alter its set point, but this is now considered
unlikely. Although the bacterial endotoxin, if injected directly into
the hypothalamic region of the brain, does indeed produce a fever, it
does so only after a time lag of an hour or two, during which time
numerous leucocytes have congregated around the point of injection, presumably producing the pyrogen. Furthermore, persons and
experimental animals which have become leucopenic (having reduced number of leucocytes) fail to produce a fever, or produce only
a slight fever, in response to an infection, another observation sug-
gesting that leucocytes are necessary for the production of the pyrogenic substance.

INTERMEDIATE CHEMICALS AND NEUROTRANSMITTERS

There is good evidence that the substances acting on the hypothalamus to alter its set point are metabolic products of arachidonic acid,
a fatty acid derived from phospholipids. The suggestion is made that
one of the actions of the endogenous pyrogen, which does not itself
cross the blood–brain barrier, is to stimulate metabolism of arachidonic acid (perhaps in certain cells at the blood–brain barrier), increasing the production of the pyrogenic compounds. Among these
compounds are prostaglandins of the E series, E_1 and E_2. Certainly
these prostaglandins, if extracted and prepared in pure form and
injected in minute quantities into the cerebral ventricles of an animal, become attached to the hypothalamic region and cause a fever.
Part of the evidence implicating arachidonic acid and its products as
agents in fevers comes from the study of the effect of the antipyretic
(fever-reducing) drugs of the aspirin type: sodium salicylate, acetyl

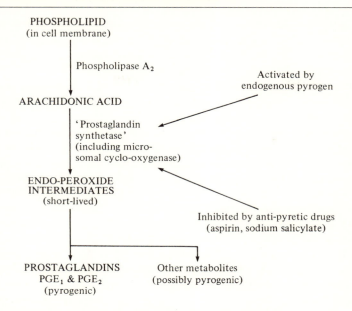

Fig. 10.2. Metabolism of arachidonic acid in production of prostaglandins.

salicylate and indomethacin. These drugs have been known for many years to reduce the body temperature in a fever. Salicylic acid can be made from the bark of willow trees, *Salix* sp., and infusions of willow bark were used in early times to reduce the fever in malaria. *In vitro* experiments have shown that these drugs inhibit the synthesis of prostaglandins from arachidonic acid, perhaps by binding irreversibly to 'prostaglandin synthetase'. Arachidonic acid, when injected into the ventricles of a rabbit's brain, rapidly causes a fever, but if indomethacin (one of the aspirin-like drugs) is injected simultaneously the effect of the arachidonic acid is abolished and the animal's temperature remains normal. This result is to be expected if the indomethacin is inhibiting production of the pyrogenic agent from the arachidonic acid. Fig. 10.2 shows these possible relationships.

However, other derivatives of arachidonic acid, besides the prostaglandins, may also be involved. One reason for thinking so is that prostaglandin inhibitors (which inhibit actions but not synthesis of prostaglandins) do not prevent the fever resulting from arachidonic acid injections into the ventricles, as they would do if prostaglandins were the only pyrogenic products involved. Another reason is that fever can still be produced after electrolytic destruction of an animal's hypothalamus, which is apparently the only region of the brain where prostaglandins are pyrogenic. So the precise nature and mode of action of derivatives of arachidonic acid in fevers has not yet been defined.

Numerous chemical transmitters are involved in neuronal pathways in the central nervous system, including the pathways controlling body temperature. Acetylcholine, catecholamines, 5-hydroxytryptamine and gamma-aminobutyric acid are the best known, and 40 or more substances have also been identified as genuine or possible transmitters in the brain. It is difficult to discover which particular transmitters are used along the neuronal pathways effecting the sensing and control of temperature, or to identify the point at which prostaglandins (or other substances produced by endogenous pyrogen) might interfere with such transmitters in fever. Experiments involving injection of possible transmitters or transmitter-blockers into or near the hypothalamus and observing consequent changes in core temperature have on the whole been uninformative and inadequately controlled. However, a well-controlled series of experiments has elucidated for the brain of one species, the sheep, the neuronal pathways and their transmitters between sensory and effector systems of thermoregulation. This mapping of the pathways was subsequently used (Bligh, Silver, Bacon and Smith, 1978; Maskrey, 1971) to point to the site of action of a pyrogen. This study, on the sheep, involved intravenous injection of typhoid–paratyphoid vaccine as a bacterial endotoxin, and the intracranial injection of acetylcholine (or its analogue) and its blocker, atropine. The results indicated (1) that at least one cholinergic synapse was involved, and (2) that the action of the endogenous pyrogen was likely to be at or near the central cold sensors but before the cholinergic synapse. When more is known in a number of species about the transmitters along the temperature-controlling pathways and their disturbance in fever, it may be possible to produce drugs which attenuate fever by blocking the central action of the pyrogen.

PHYSIOLOGICAL SIGNIFICANCE OF FEVER

It has been suggested from time to time that the fever accompanying an infection was of some biological advantage to the sufferer, perhaps by killing or reducing the growth of the invading microorganism. The fever, on this view, would represent part of the body's defence mechanism against the infection. This idea has been supported by the observation that certain poikilothermic species (lizards, amphibia, fish) survive infections better in a hot environment which allows their body temperature to rise to 42 °C than in a cooler environment. The view that fever is beneficial to the sufferer implies that the growth or multiplication of some pathogenic microorganisms is impeded by the raised body temperature alone, or as part of a more complex response of the body to the infection. It is true that certain bacteria (gonococci and spirochaetes), which are not strongly pyrogenic, do not survive *in vitro* at a temperature of 41–42 °C; indeed raising the host's body temperature has been used in the treatment of infections with these particular organisms. However, most pathogenic organisms survive and even thrive at this temperature, so a raised body temperature would not alone be expected to help the sufferer. Again, by a direct Q_{10} effect the high body temperature might be expected to increase all activities including those of the host's phagocytes combating the invasion, but a similar rise in the physiological and biochemical activities of the invading microorganisms, including that of multiplication, would also be expected. Thus there is little reason to suppose that the raised temperature *per se* would preferentially favour the host over the microorganism.

There is in fact a good argument *against* the idea that the fever is in some way beneficial to the sufferer. For many years antipyretic drugs have been used to help to lower a person's body temperature in infectious disease. If the fever were an advantage in combating the infection, such a treatment would be expected to worsen the outcome of the disease; but such a deleterious effect of lowering body temperature by drugs has not been shown convincingly. The question of whether, for homeotherms, the fever of infection is harmful, or beneficial, or neither, is at present unsettled. There is evidence that the production of essential iron-transport compounds (siderophores) in some microorganisms is lowered by a rise in temperature, the organisms becoming less viable and less able to multiply in consequence of this reduction. In those (poikilothermic) species in which hyperthermia (a raised preferred temperature) appears to be beneficial in combating the disease, the effect of temperature on siderophore production might be the means by which the benefit is gained.

OTHER EFFECTS OF INFECTION

Even though the hyperthermia of infection may, alone, be of no advantage to the (homeothermic) host, its occurrence together with other responses to the infection may be beneficial. One such process is the fall in the level of plasma iron and zinc. Since many species of bacteria grow poorly *in vitro* in an iron-deficient medium (and grow well in one which is iron-enriched) it has been suggested that this reduction in the host's plasma iron might reduce the growth of pathogenic organisms *in vivo*. A leucocyte-derived protein, the *leucocyte endogenous mediator* which may or may not be different from the endogenous pyrogen also produced from leucocytes, may be responsible for the decrease in plasma iron, as well as for changes in the levels of other trace metals in plasma. Although it is now considered likely that the leucocyte endogenous mediator and the leucocyte pyrogen are one substance with two distinct and independent effects, these effects can be separated: infection of new-born mammals produces the expected fall in plasma iron but with little or no fever, and it has been shown recently that while indomethacin attenuates the fever produced by endogenous pyrogen, it does not attenuate the effect of endogenous pyrogen on plasma iron and zinc. Thus the latter effect apparently does not involve arachidonic acid metabolism as does fever. In this connection, there may well prove to be a chemical link-up between the hyperthermia of fever and that of exercise. A protein similar to or identical with endogenous pyrogen has been extracted from plasma and leucocytes of blood samples taken from humans immediately after the subjects had completed an hour's vigorous exercise. This material, when injected into rats, produced not only a rise in their core temperature but also a reduction of their plasma iron and zinc (Cannon and Kluger, 1983). It is of interest that the two types of hyperthermia, of very different origin, seem to have some biochemical features in common.

The relationships of the various chemicals and processes that may be involved in fevers are shown diagrammatically in Fig. 10.3. It must be emphasized that this diagram is based on a synthesis of pieces of evidence and will require modification as knowledge increases.

MAXIMUM FEVER

The new set point temperature present during a fever may be below 41 °C (106 °F) but is never above it. If a human's body temperature

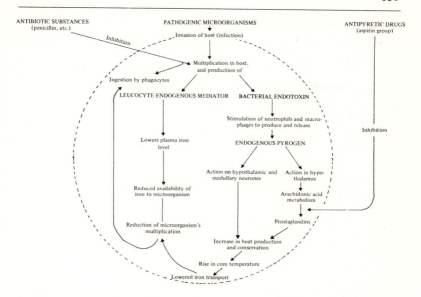

Fig. 10.3. Diagram showing some of the factors which may be involved in the initiation of and recovery from a fever due to infection. Dotted line indicates the limits of the host's body.

does rise above 41 °C, this is the start of an uncontrollable runaway hyperthermia from which his own unaided physiological processes cannot rescue him, and from which he can be saved only by artificial aids such as immersion in cool water or sponging beneath a fan to increase heat loss. The reason why 41 °C core temperature represents the upper limit, for humans, is unknown.

Humans (and animals) usually drink abundantly during a fever but eat little. In consequence, the increased metabolism with the rise of body temperature must use, for fuel, the body's own tissue stores of energy. So after recovery from a severe infection, the sufferer is nearly always found to have lost weight. He may also have lost protein as a result of the muscular wasting associated with rest in bed. This would be shown by negative nitrogen balance during the infection.

SUMMARY

The characteristic of fever is that elevated core temperature is not countered by corrective thermoregulatory processes, as are other forms of hyperthermia. Whatever central process sets the level of body temperature has been temporarily changed in such a way that the set point is raised. This organized elevation of core temperature is usually indicative of a bacterial or viral infection; but a set of symptoms indistinguishable from fever can occur in the absence of infection and may be indicative of a cancerous tissue.

The primary causative factor in infection-related fever is a heat-stable lipopolysaccharide product of the microorganism which has toxic effects on many of the host's tissues and systems. This substance is called bacterial endotoxin, though viruses may produce a similar substance. The invading microorganisms are phagocytosed by the phagocytic cells of the host's reticulo-endothelial system. In addition to engulfing and destroying the microorganism, the phagocytes produce (1) a small heat-labile protein, the endogenous (or leucocytic) pyrogen, which acts directly or indirectly on the central nervous system's structures to cause fever; and (2) a leucocyte endogenous mediator (possibly identical with (1)) which reduces the iron and zinc concentrations in the host's plasma. The endogenous pyrogen may have a direct effect on the central neurones involved in thermoregulation, or it may act through one or more intermediary chemical processes. One such process is probably the stimulation of arachidonic acid metabolism, leading to the production of prostaglandins and other pyrogenic chemicals.

There is a limit to the extent of rise in core temperature in fever. Core temperature seldom exceeds 41 °C in man. A rise above 41 °C indicates a failure which causes the body temperature to go out of control and rise rapidly; heat death is then inevitable unless physical (environmental) methods of extracting heat from the body are applied promptly.

Appendix

UNITS: EQUIVALENTS IN SYSTÈME INTERNATIONALE (SI), C.G.S. AND THE BRITISH SYSTEM

	S.I.	c.g.s.	British
Length, area, and volume	1 m	100 cm	3.281 ft
	0.3048 m	30.48 cm	1 ft
	1 km		0.6214 mile
	1.609 km		1 mile
	1 m^2	10^4 cm^2	10.76 ft^2
	0.0929 m^2	929.0 cm^2	1 ft^2
	1 m^3	10^6 cm^3	35.31 ft^3
	0.02832 m^3	28.32 × 10^3 cm^3	1 ft^3
Velocity	1 m s^{-1}	100 cm s^{-1}	2.237 miles h^{-1} = 196.9 ft min^{-1} = 3.281 ft s^{-1}
	0.4470 m s^{-1}	44.70 cm s^{-1}	1 mile h^{-1} = 88 ft min^{-1} = 1.467 ft s^{-1}
Mass	1 kg	10^3 g	2.205 lb
	0.4536 kg	453.6 g	1 lb
Pressure	1 Pascal (Pa) = 1 N m^{-2}	10 g cm^{-1} s^{-2} = 10^{-2} mbar = 7.501 × 10^{-3} mmHg	0.02089 lb force ft^{-2}
	47.88 Pa	478.8 g cm^{-1} s^{-2} = 0.4788 mbar = 0.3591 mmHg	1 lb force ft^{-2}
	1 atmosphere 101.3 kPa	= 1013 mbar = 760.0 mmHg = 760.0 Torr	2116 lb force ft^{-2} = 14.69 lb force in^{-2}
Temperature	1°C (or K)	1°C (or K)	1.8°F
	0.5556°C	0.5556°C	1°F
Heat	1 J	0.2390 calorie (cal)	0.9485 × 10^{-3} BTU
	4.184 J	1 cal (thermochemical)	3.968 × 10^{-3} BTU
	1.054 kJ	0.2520 kcal	1 BTU

Metabolic rate	$1 \text{ W} = 1 \text{ J s}^{-1}$	$0.8604 \text{ kcal h}^{-1}$	3.414 BTU h^{-1}
	1.162 W	1 kcal h^{-1}	3.968 BTU h^{-1}
	0.2929 W	$0.2520 \text{ kcal h}^{-1}$	1 BTU h^{-1}
	1 MJ day^{-1} Ĺ28.00.2390 Mcal day^{-1}		$0.9485 \times 10^3 \text{ BTU day}^{-1}$
	$= 11.57 \text{ W}$		
	$4.184 \text{ MJ day}^{-1}$	1 Mcal day^{-1}	$3.968 \times /10^3 \text{ BTU day}^{-1}$
	$= 48.43 \text{ W}$		
	$1.054 \text{ kJ day}^{-1}$	$0.2520 \text{ kcal day}^{-1}$	1 BTU day^{-1}
Thermal insulation	$1^\circ\text{C m}^2 \text{ W}^{-1}$	$1.162^\circ\text{C m}^2 \text{ h kcal}^{-1}$	$5.674 \text{ F ft}^2 \text{ h BTU}^{-1}$
	$0.8604^\circ\text{C m}^2 \text{ W}^{-1}$	$1^\circ\text{C m}^2 \text{ h kcal}^{-1}$	$4.882 \text{ F ft}^2 \text{ h BTU}^{-1}$
	$0.1762^\circ\text{C m}^2 \text{ W}^{-1}$	$0.2048^\circ\text{C m}^2 \text{ h kcal}^{-1}$	$1 \text{ F ft}^2 \text{ h BTU}^{-1}$
	$1^\circ\text{C m}^2 \text{ day MJ}^{-1}$	$4.184^\circ\text{C m}^2 \text{ day Mcal}^{-1}$	$0.4903 \text{ F ft}^2 \text{ h BTU}^{-1}$
	$= 0.0864^\circ\text{C m}^2 \text{ W}^{-1}$		
	$0.2390^\circ\text{C m}^2 \text{ day MJ}^{-1}$	$1^\circ\text{C m}^2 \text{ day Mcal}^{-1}$	$0.1172 \text{ F ft}^2 \text{ h BTU}^{-1}$
	$2.040^\circ\text{C m}^2 \text{ day MJ}^{-1}$	$8.534^\circ\text{C m}^2 \text{ day Mcal}^{-1}$	$1 \text{ F ft}^2 \text{ h BTU}^{-1}$

Note: although care has been taken in the calculation and presentation of these values, the authors and publisher do not accept liability for any loss or damage suffered as a result of their application and use.

(From Mount, 1979)

Suggestions for further reading

CHAPTER ONE

Dauncey, M.J., Murgatroyd, P.A. & Cole, T.J. (1979). A human calorimeter for the direct and indirect measurement of energy expenditure. *Br. J. Nutr.* **39**, 557–61.

CHAPTER TWO

Kleiber, M. (1975). *The Fire of Life: An Introduction to Animal Energetics*, 2nd edn. New York: John Wiley.

CHAPTER THREE

Edholm, O.G. & Weiner, J.S. (1981). Thermal physiology. In *Principles and Practice of Human Physiology*, ed. O.G. Edholm & J.S. Weiner, pp. 111–90. London: Academic Press.
Gordon, M.S., Bartholomew, G.A., Grinnell, A.D., Jørgensen, C.B. & White, F.N. (1982). *Animal function: Principles and Adaptations*, 4th edn. New York: Macmillan.

CHAPTER FOUR

Monteith, J.L. & Mount, L.E. (eds.) (1974). *Heat Loss from Animals and Man: Assessment and Control*. London: Butterworth.

CHAPTER FIVE

Bligh, J. (1973). *Temperature Regulation in Mammals and other vertebrates*. Amsterdam: North Holland Publishing Co.
Gordon, C.J. & Heath, J.E. (1983). Reassessment of the neural control of body temperature: importance of oscillating neural and motor components. *Comp. Biochem. Physiol.* **74A**, 479–89.
Hardy, J.D., Gagge, A.P. & Stolwijk, J.A.J. (eds.) (1970). *Physiological and Behavioural Thermoregulation*. Springfield: Charles C. Thomas.

Hensel, H. (1981). *Thermoreception and Temperature Regulation.* London: Academic Press.

Whittow, G.C. (ed.) (1970). *Comparative Physiology of Thermoregulation*, Vols. II and III. London: Academic Press.

CHAPTER SIX

Fox, R.H., MacGibbon, R., Davies, L. & Woodward, P.M. (1973). Problems of the old and the cold. *Br. med. J.* **1**, 21–4.

CHAPTER SEVEN

Edholm, O.G. & Gunderson, E.K.E. (eds.) (1973). *Polar Human Biology.* London: Heinemann.

Folk, G.E. (1974). *Textbook of Environmental Physiology.* Philadelphia: Lea & Febiger.

Ingram, D.L. & Mount, L.E. (1975). *Man and Animals in Hot Environments.* New York: Springer.

Mount, L.E. (1979). *Adaptation to Thermal Environment: Man and his Productive Animals.* London: Edward Arnold.

CHAPTER EIGHT

Haresign, W., Swan, H. & Lewis, D. (1977). *Nutrition and the Climatic Environment.* London: Butterworth.

CHAPTER NINE

Åstrand, P.-O. & Rodahl, K. (1977). *Textbook of Work Physiology: Physiological Bases of Exercise*, 2nd edn. New York: McGraw-Hill.

Bassey, E.J. & Fentem, P.H. (1981). Work physiology. In *Principles and Practice of Human Physiology*, ed. O.G. Edholm & J.S. Weiner, pp. 19–110. London: Academic Press.

CHAPTER TEN

Mims, C.A. (1977). *The Pathogenesis of Infectious Disease.* London: Academic Press.

Bligh, J. (1982). Thermoregulation: its change during infection with endotoxin-producing micro-organisms. In *Handbook of Experimental Pharmacology*, vol. 60. *Pyretics and Antipyretics*, ed. A.S. Milton, pp. 25–71. Berlin: Springer-Verlag.

References

Alexander, G. (1974). Heat loss from sheep. In *Heat Loss from Animals and Man: Assessment and Control*, ed. Monteith, J.L. & Mount, L.E., pp. 173–203. London: Butterworth.

Alexander, G. (1975). Body temperature control in mammalian young. *Br. med. Bull.* **31** 61–8.

Aschoff, J. & Wever, R. (1958). Kern and Schale im Wärmehaushalt des Menschen. *Naturwissenschaften*, **45**, 477–85.

Åstrand, P-O. & Rodahl, K. (1977). *Textbook of Work Physiology: Physiological Bases of Exercise*. 2nd edn. New York: McGraw-Hill.

Bassey, E.J. & Fentem, P.H. (1981). Work physiology. In *Principles and Practice of Human Physiology*, ed. Edholm, O.G. & Weiner, J.S., pp. 19–110. London: Academic Press.

Benzinger, T.H. (1959). On physical heat regulation and the sense of temperature in man. *Proc. natn. Acad. Sci. U.S.A.* **45**, 645–59.

Bligh, J. & Harthoorn, A.M. (1965). Continuous radiotelemetric records of the deep body temperature of some unrestrained African mammals under near-natural conditions. *J. Physiol., Lond.* **176**, 145–62.

Bligh, J., Silver, A., Bacon, M.J. & Smith, C.A. (1978). The central role of a cholinergic synapse in thermoregulation in the sheep. *J. therm. Biol.* **3**, 147–51.

Burnand, E.D. & Cross, K.W. (1958). Rectal temperature in the newborn after birth asphyxia. *Br. med. J.* **2**, 1197–9.

Cabanac, M. & Caputa, M. (1979). Open loop increase in trunk temperature produced by face cooling in working humans. *J. Physiol., Lond.* **289**, 163–74.

Cabanac, M. & Massonet, B. (1977). Thermoregulatory response as a function of core temperature in humans. *J. Physiol., Lond.* **265**, 587–96.

Cairnie, A.B. and Pullar, J.D (1959). An investigation into the efficient use of time in the calorimetric measurement of heat output. *Br. J. Nutr.* **13**, 431–9.

Cannon, J.G. & Kluger, M.J. (1983) Endogenous pyrogen activity in human plasma after exercise. *Science, N.Y.* **220**, 617–19.

Cena, K. (1974). Radiative heat loss from animals and man. In *Heat Loss from Animals and Man: Assessment and Control*, ed. Monteith, J.L. & Mount, L.E., pp. 33–58. London: Butterworth.

Chatfield, P.O., Lyman, C.P. & Irving, L. (1953). Physiological adaptation to cold of peripheral nerve in the leg of the herring gull (*Larus argentatus*). *Am. J. Physiol.* **172**, 639–44.

Clark, R.P. & Toy, N. (1975). Natural convection around the human head. *J. Physiol., Lond.* **244**, 283–93.

Collins, K.J., Exton-Smith, A.N. & Doré, C. (1981). Urban hypothermia: preferred temperature and thermal perception in old age. *Br. med. J.* **1**, 175–85.

Corbett, J.C., Johnson, R.H., Krebs, H.A., Walton, J.L. & Williamson, D.H. (1969). The effect of exercise on blood ketone-body concentrations in athletes and untrained subjects. *J. Physiol., Lond.* **201**, 83–4.

Dallenbach, K.M. (1927). The temperature spots and end-organs. *Am. J. Psychol.* **39**, 402–27.

Dauncey, M.J. (1980). Metabolic effects of altering the 24 h energy intake in man, using direct and indirect calorimetry. *Br. J. Nutr.* **43**, 257–69.

Dauncey, M.J. & Bingham, S.A. (1983). Dependence of 24 h energy expenditure in man on the composition of the nutrient intake. *Br. J. Nutr.* **50**, 1–13.

Dauncey, M.J., Ingram, D.L., Walters, E. & Legge, K.F. (1983). Evaluation of separate and combined effects of environmental temperature and nutrition on growth and development. *J. agric. Sci., Camb.* (in press).

Durnin, J.V.G.A. & Rahaman, M.M. (1967). An assessment of the amount of fat in the human body from measurements of skinfold thickness. *Br. J. Nutr.* **21**, 681–9.

Durnin, J.V.G.A. & Womersley, J. (1974). Body fat assessed from total body density and its estimation from skinfold thickness: measurements on 481 men and women aged from 16 to 72 years. *Br. J. Nutr.* **32**, 77–97.

Edholm, O.G., Bedford, T., Ellis, F.P. & Mackworth, N.H (eds.) (1960). *Physiological Responses to Hot Environments.* Medical Research Council Special Report Series 298.

Edholm, O.G., Fletcher, J.G., Widdowson, E.M. & McCance, R.A. (1955). The energy expenditure and food intake of individual man. *Br. J. Nutr.* **9**, 286–300.

Edholm, O.G. & Weiner, J.S. (eds.) (1981). *Principles and Practice of Human Physiology.* London: Academic Press.

Folk, G.E. (1974). *Textbook of Environmental Physiology.* Philadelphia: Lea & Febiger.

Fox, R.H., Woodward, P.M., Exton-Smith, A.N., Green, M.F., Donnison, D.V. & Wicks, M.H. (1973). Body temperatures in the elderly: a national study of physiological, social and environmental conditions. *Br. med. J.* **2**, 200–06.

Gordon, M.S., Bartholomew, G.A., Grinnell, A.D., Jørgensen, C.B. & White, F.N. (1968). *Animal Function: Principles and Adaptations.* New York and London: MacMillan.

Graham, N.M., Wainman, F.W., Blaxter, K.L. & Armstrong, D.G. (1959). Environmental temperature, energy metabolism and heat regulation in sheep. *J. agric. Sci., Camb.* **52**, 13–24.

Guyton, A.C. (1975). *Textbook of Medical Physiology*, 5th edn. Philadelphia: Saunders.

Hardy, R.N. (1979). *Temperature and Animal Life*, 2nd edn. Studies in Biology No. 35. London: Edward Arnold.

Hey, E.N. (1974). Physiological control over body temperature. In *Heat*

Loss from Animals and Man: Assessment and Control, ed. Monteith, J.L. & Mount, L.E., pp.77–95. London: Butterworth.

Hey, E.N. & Mount, L.E. (1967). Heat loss from babies in incubators. *Arch. Dis. Childh.* **42**, 75–84.

Hey, E.N. & O'Connell, B. (1970). Oxygen consumption and heat balance in the cot-nursed baby. *Arch. Dis. Childh.* **45**, 335–43.

Hill, J.R. & Rahimtulla, K.A. (1965). Heat balance and the metabolic rate of newborn babies in relation to environmental temperature: the effects of age and of weight on the basal metabolic rate. *J. Physiol., London.* **180**, 239–65.

Ingram, D.L. & Mount, L.E. (1975). *Man and Animals in Hot Environments*. New York: Springer.

Irving, L. (1964). Terrestrial animals in cold: birds and mammals. In *Adaptation to the Environment*, ed. Dill, D.B. Handbook of Physiology, Section 4. Washington D.C: American Physiological Society.

Jessen, C. (1977). Interaction of air temperature and core temperature in thermoregulation of the goat. *J. Physiol., Lond.* **264**, 585–606.

Kleiber, M. (1947) Body size and metabolic rate. *Physiol. Rev.* **27**, 511–41.

Knibs, A.V. (1971). Some physiological effects of intensity and frequency of exercise in young non-athletic females. *J. Physiol., Lond.* **216**, 25–6P.

Lusk, G. (1928). *The Elements of the Science of Nutrition*, 4th edn. Philadelphia: Saunders.

Lyman, C.P. (1948). The oxygen consumption and temperature regulation of hibernating hamsters. *J. exp. Zool.* **109**, 55–78.

Maskrey, M. (1971). 'Transmitters and modulator substances in the control of body temperature'. Unpublished Ph.D.thesis, University of Cambridge.

Mayer, J. & Bullen, B. (1960). Nutrition and athletic performance. *Physiol. Rev.* **40**, 369–97.

Mount, L.E. (1960). The influence of huddling and body size on the metabolic rate of the young pig. *J. agric. Sci., Camb.* **55**, 101–05.

Mount, L.E. (1979). *Adaptation to Thermal Environment: Man and his Productive Animals*. London: Edward Arnold.

Newland, H.W., McMillen, W.N. & Reineke, E.P. (1952). Temperature adaptation in the baby pig. *J. Anim. Sci.* **11**, 118–33.

Pugh, L.G.C. & Edholm, O.G. (1955). The physiology of channel swimmers. *Lancet*, **2**, 761–8.

Schmidt-Nielsen, K., Schmidt-Nielsen, B., Jarnum, S.A. & Houpt, T.R. (1957). Body temperature of the camel and its relation to water economy. *Am. J. Physiol.* **188**, 103–12.

Scholander, P.F., Walters, V., Hock, R. & Irving, L. (1950). Body insulation of some arctic and tropical mammals and birds. *Biol. Bull. mar. biol. Lab., Woods Hole*, **99**, 225–36.

Setchell, B.P. (1978). *The mammalian testis*. London: Elek.

Veghte, J.H. & Herreid, C.F. (1965). Radiometric determination of feather insulation and metabolism of arctic birds. *Physiol. Zool.* **38**, 267–75.

Whittow, G.C. (1962). The significance of the extremities of the ox (*Bos taurus*) in thermoregulation. *J. agric. Sci., Camb.* **58**, 109–20.

Whittow, G.C. (ed.) (1971). *Comparative Physiology of Thermoregulation*, Vols. I and II. London: Academic Press.

Zotterman, Y. (1953). Special senses: thermal receptors. *A. Rev. Physiol.* **15**, 357–72.

Index